Real-Time Embedded Systems

Quantitative Software Engineering Series

Lawrence Bernstein, Series Editor

The Quantitative Software Engineering Series focuses on the convergence of systems engineering and software engineering with emphasis on quantitative engineering trade-off analysis. Each title brings the principles and theory of programming in the large and industrial strength software into focus. This practical series helps software developers, software engineers, systems engineers, and graduate students understand and benefit from this convergence through the unique weaving of software engineering case histories, quantitative analysis, and technology into the project effort. You will find that each publication reinforces the series goal of assisting the reader with producing useful, well-engineered software systems.

Software Testing: Concepts and Operations
   by Ali Mili, Fairouz Tchier

Enterprise Software Architecture and Design: Entities, Services, and Resources
   by Dominic Duggan

Oracle Database Performance and Scalability: A Quantitative Approach
   by Henry H. Liu

Trustworthy Compilers
   by Vladimir O. Safonov

Managing the Development of Software-Intensive Systems
   by James McDonald

Software Performance and Scalability: A Quantitative Approach
   by Henry H. Liu

Web Application Design and Implementation: Apache 2, PHP5, MySQL, JavaScript, and Linux/UNIX
   by Steven A. Gabarro

Software Measurement and Estimation: A Practical Approach
   by Linda M. Laird, M. Carol Brennan

Trustworthy Systems through Quantitative Software Engineering
   by Lawrence Bernstein, C. M. Yuhas

Real-Time Embedded Systems
   by Jiacun Wang

Real-Time Embedded Systems

Jiacun Wang

The right of Jiacun Wang to be identified as the author of this work has been asserted in accordance with law.

Registered Office
John Wiley & Sons, Inc., 111 River Street, Hoboken, NJ 07030, USA

Editorial Office
111 River Street, Hoboken, NJ 07030, USA

For details of our global editorial offices, customer services, and more information about Wiley products visit us at www.wiley.com.

Wiley also publishes its books in a variety of electronic formats and by print-on-demand. Some content that appears in standard print versions of this book may not be available in other formats.

Library of Congress Cataloging-in-Publication Data

Names: Wang, Jiacun, 1963- author.
Title: Real-time embedded systems / by Jiacun Wang.
Description: Hoboken, NJ, USA : Wiley, 2017. | Series: Quantitative software engineering series | Includes bibliographical references and index. | Identifiers: LCCN 2017015038 (print) | LCCN 2017030269 (ebook) | ISBN 9781119420705 (pdf) | ISBN 9781119420682 (epub) | ISBN 9781118116173 (hardback)
Subjects: LCSH: Embedded computer systems. | Real-time data processing. | BISAC: TECHNOLOGY & ENGINEERING / Electronics / Microelectronics.
Classification: LCC TK7895.E42 (ebook) | LCC TK7895.E42 W36 2017 (print) | DDC 006.2/2–dc23
LC record available at https://lccn.loc.gov/2017015038

Cover image: © spainter_vfx/Gettyimages
Cover design by Wiley

Set in 10/12pt Warnock by SPi Global, Chennai, India

Printed in the United States of America

10 9 8 7 6 5 4 3 2 1

Contents

Preface

Real-time embedded systems play a significant role in our daily life. These systems are inside our cars, cell phones, and home electrical appliances. Industrial process control, telecommunication, signal processing, vehicle navigation, air traffic control, and space exploration all depend on real-time embedded system technologies. The real-time embedded application market has been further driven to a new high by recent advances in the information and communication technology and by the emergence of Internet of things, ubiquitous computing, and pervasive computing. Therefore, there is an increasing demand from the related branches of industry for computer scientists and software engineers who are particularly capable of real-time embedded system hardware and software design and development. This textbook aims to prepare students with the fundamental knowledge and skills that are needed to meet the challenges.

This book introduces the characteristics of real-time embedded systems, typical embedded hardware components, fundamental real-time operating system features, well-known real-time task scheduling algorithms, and widely used resource access control protocols. It also presents several formal approaches for real-time embedded system design, modeling, analysis and critical property verification. For those who are interested in real-time software development, the text will familiarize them with techniques and skills in concurrent programming and real-time task implementation.

Monmouth University
West Long Branch, NJ
January 15, 2017

Book Layout

This book has four parts. Part 1 includes Chapters 1–3, which introduces the fundamental concepts and characteristics of real-time embedded systems, embedded system hardware basics, general operating systems, and real-time operating systems. The automobile antilock braking system is used as an example to show the components and characteristics of real-world real-time embedded systems.

Part 2 includes Chapters 4–6, focusing on real-time system scheduling, task assignment, resource access control, and real-time embedded system programming. Clock-driven and priority-driven scheduling algorithms, as well as several task assignment algorithms, are discussed in Chapter 4. Resource sharing issues and some well-known resource access control protocols that tackle the priority inversion and deadlock problems are presented in Chapter 5. Real-time task implementation, intertask synchronization and communication, and concurrent programming details are discussed in Chapter 6.

Part 3 is composed of Chapters 7–10, which introduces various modeling and analysis techniques for real-time embedded systems. Among them, finite-state machines are a traditional model of computation good for logic circuits and software design and are introduced in Chapter 7. UML state machines are an extension to traditional finite-state machines with hierarchy and orthogonality and composed of a rich set of graphical notations. They are introduced in Chapter 8. Petri nets are a high-level model, which is very powerful for event-driven system modeling and analysis. Timed Petri nets allow users to verify systems' timing constraints. The theory and applications of Petri nets and timed Petri nets are presented in Chapter 9. Model checking is a technique that verifies system properties against a model using a software tool. Model checking principles, the NuSMV model checker, and its underlying temporal logic and description language are introduced in Chapter 10.

Chapter 11 alone is the final part of the book, which briefly discusses some practical issues with real-time embedded system design and development, including software reliability, aging, security, safety, and power consumption.

Audience

This book is primarily written as a textbook for undergraduate or graduate courses in computer engineering, software engineering, computer science, and information technology programs in the subject of embedded and real-time systems. For students with certain level of familiarity with computer programming, operating systems, and computer architecture, the book will extend their knowledge and skills into the area of real-time embedded computing that has profound influence on the quality of our daily life.

Portions of the book could be used as reading material for undergraduate computer engineering, computer science, and software engineering capstone courses or as a reference for students who are doing research in the area of real-time embedded systems.

The book is also useful to industrial practitioners with real-time and embedded software design and development responsibilities.

Acknowledgments

This project was partially supported by Mexico's CONACYT program "Estancias Sabáticas en México para Extranjeros para la Consolidación de Investigación." Chapter 9 was written by Dr Xiaoou Li, Professor of CINVESTAV-IPN, Mexico. In writing the book, I received constant help from my colleague and friend over a decade, Dr William Tepfenhart, Professor of Monmouth University. I discussed the contents of Chapters 6–8 intensively with him, and his insights into the subjects greatly influenced the writing of the three chapters. Dr Xuemin Chen, Professor of Texas Southern University, reviewed the book proposal and offered good suggestions on the layout of the book and contents of the first two chapters. Professor Lijun Chen, Xi'an University of Posts and Telecommunication, reviewed the draft of Chapter 6 and provided her constructive comments. Mr Bin Hu, a master's degree student of Monmouth University, tested most of the programming projects included in the book. I thank all of them for their generous help.

1

Introduction to Real-Time Embedded Systems

Real-time embedded systems have become pervasive. They are in your cars, cell phones, Personal Digital Assistants (PDAs), watches, televisions, and home electrical appliances. There are also larger and more complex real-time embedded systems, such as air-traffic control systems, industrial process control systems, networked multimedia systems, and real-time database applications. It is reported that in the Lexus LS-460 released in September 2006, there are more than 100 microprocessors embedded when all optional features are installed. It is also estimated that 98% of all microprocessors are manufactured as components of embedded systems. In fact, our daily life has become more and more dependent on real-time embedded applications. This chapter explains the concepts of embedded systems and real-time systems, introduces the fundamental characteristics of real-time embedded systems, and defines hard and soft real-time systems. The automotive antilock braking system (ABS) is used as an example to show a real-world embedded system.

1.1   Real-Time Embedded Systems

An *embedded system* is a microcomputer system embedded in a larger system and designed for one or two dedicated services. It is embedded as part of a complete device that often has hardware and mechanical parts. Examples include the controllers built inside our home electrical appliances. Most embedded systems have real-time computing constraints. Therefore, they are also called real-time embedded systems. Compared with general-purpose computing systems that have multiple functionalities, embedded systems are often dedicated to specific tasks. For example, the embedded airbag control system is only responsible for detecting collision and inflating the airbag when necessary, and the embedded controller in an air conditioner is only responsible for monitoring and regulating the temperature of a room.

Real-Time Embedded Systems, First Edition. Jiacun Wang.
© 2017 John Wiley & Sons, Inc. Published 2017 by John Wiley & Sons, Inc.

Another noteworthy difference between a general-purpose computing system and an embedded system is that a general-purpose system has full-scale operating system support, while embedded systems may or may not have operating system support at all. Many small-sized embedded systems are designed to perform simple tasks and thus do not need operating system support.

Embedded systems are *reactive systems* in nature. They are basically designed to regulate a physical variable in response to the input signal provided by the end users or sensors, which are connected to the input ports. For example, the goal of a grain-roasting embedded system is regulating the temperature of a furnace by adjusting the amount of fuel being injected into the furnace. The regulation or control is performed based on the difference between the desired temperature and the real temperature detected by temperature sensors.

Embedded systems can be classified based on their complexity and performance into small-scale, medium-scale, and large-scale. Small-scale systems perform simple functions and are usually built around low-end 8- or 16-bit microprocessors or microcontrollers. For developing embedded software for small-scale embedded systems, the main programming tools are an editor, assembler, cross-assembler, and integrated development environment (IDE). Examples of small-scale embedded systems are mouse and TV remote control. They typically operate on battery. Normally, no operating system is found in such systems.

Medium-scale systems have both hardware and software complexities. They use 16- or 32-bit microprocessors or microcontrollers. For developing embedded software for medium-scale embedded systems, the main programming tools are C, C++, JAVA, Visual C++, debugger, source-code engineering tool, simulator, and IDE. They typically have operating system support. Examples of medium-scale embedded systems are vending machines and washing machines.

Large-scale or sophisticated embedded systems have enormous hardware and software complexities, which are built around 32- or 64-bit microprocessors or microcontrollers, along with a range of other high-speed integrated circuits. They are used for cutting-edge applications that need hardware and software codesign techniques. Examples of large-scale embedded systems are flight-landing gear systems, car braking systems, and military applications.

Embedded systems can be *non-real-time* or *real-time*. For a non-real-time system, we say that it is correctly designed and developed if it delivers the desired functions upon receiving external stimuli or internal triggers, with a satisfied degree of QoS (Quality of Service). Examples are TV remote controls and calculators.

Real-time systems, however, are required to compute and deliver correct results within a specified period of time. In other words, a job of a real-time system has a *deadline*, being it hard or soft. If a hard deadline is missed, then the result is useless, even if it is correct. Consider the airbag control system in automobiles. Airbags are generally designed to inflate in the cases of frontal impacts in automobiles. Because vehicles change speed so quickly in a crash, airbags must inflate rapidly to reduce the risk of the occupant hitting the vehicle's interior. Normally, from the onset of the crash, the entire deployment and inflation process is about 0.04 seconds, while the limit is 0.1 seconds.

Non-real-time embedded systems may have time constraints as well. Imagining if it takes more than 5 seconds for your TV remote control to send a control signal to your TV and then the embedded device inside the TV takes another 5 seconds to change the channel for you, you will certainly complain. It is reasonable that consumers expect a TV to respond to remote control event within 1 second. However, this kind of constraints is only a measure of system performance.

Traditional application domains of real-time embedded systems include automotive, avionics, industrial process control, digital signal processing, multimedia, and real-time databases. However, with the continuing rapid advance in information and communication technology and the emergence of Internet of things and pervasive computing, real-time embedded applications will be found in any objects that can be made smart.

1.2 Example: Automobile Antilock Braking System

A distinguished application area of real-time embedded systems is automobiles. Automotive embedded systems are designed and developed to control engine, automatic transmission, steering, brake, suspension, exhaustion, and so on. They are also used in body electronics, such as instrument panel, key, door, window, lighting, air bag, and seat bag. This section introduces the ABS.

An ABS is an automobile safety system. It is designed to prevent the wheels of a vehicle from locking when brake pedal pressure is applied, which may occur all of a sudden in case of emergency or short stopping distance. A sudden lock of wheels will cause moving vehicles to lose tractive contact with the road surface and skid uncontrollably. For this reason, ABS also stands for *antiskid braking system*. The main benefit from ABS operation is retaining the directional control of the vehicle during heavy braking in rare circumstances.

1.2.1 Slip Rate and Brake Force

When the brake pedal is depressed during driving, the wheel velocity (the tangential speed of the tire surface) decreases, and so does the vehicle velocity. The decrease in the vehicle velocity, however, is not always synchronized with the wheel velocity. When the maximum friction between a tire and the road surface is reached, further increase in brake pressure will not increase the *braking force*, which is the product of the weight of the vehicle and the friction

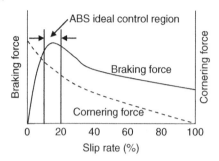

Figure 1.1 Relationship among the slip rate, braking force, and cornering force.

coefficient. As a consequence, the vehicle velocity is greater than the wheel speed, and the wheel starts to skid. The wheel slip rate, s, is defined as

$$s = \frac{V - \omega R}{V}$$

where V, ω, and R denote the vehicle speed, wheel angular velocity, and wheel rolling radius, respectively. Under normal driving conditions, $V = \omega R$ and thus, $s = 0$. During severe braking, it is common to have $\omega = 0$ and $V > 0$, which translates to $s = 1$, a case of wheel lockup.

Figure 1.1 describes the relationship among the slip rate, braking force, and cornering force. The continuous line in the figure represents the relationship between the slip rate and braking force. It shows that the braking force is the largest when the slip rate is between 10% and 20%. The braking distance is the shortest at this rate. With further increase in slip rate, the braking force is decreased, which results in a longer braking distance. The dashed line depicts the relationship between the slip rate and *cornering force*. The cornering force is generated by a vehicle tire during cornering. It works on the front wheels as steering force and on the rear wheels to keep the vehicle stable. It decreases as the slip rate increases. In case of a lockup, the cornering force becomes 0 and steering is disabled.

1.2.2 ABS Components

The ABS is composed of four components. They are speed sensors, valves, pumps, and an electrical control unit (ECU). Valves and pumps are often housed in hydraulic control units (HCUs).

1.2.2.1 Sensors
There are multiple types of sensors used in the ABS. A *wheel speed sensor* is a sender device used for reading a vehicle's wheel rotation rate. It is an electromagnet illustrated in Figure 1.2. As the sensor rotor rotates, it induces AC voltage in the coil of the electromagnet. When the rotation of the sensor rotor increases, the magnitude and frequency of induced voltage increase as well.

Figure 1.2 Wheel speed sensor.

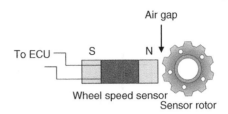

A *deceleration sensor* measures the vehicle's rate of deceleration. It is a switch type of sensor. It uses phototransistors that can be activated by light. In a deceleration sensor, two Light-Emitting Diode (LEDs) aim at two phototransistors that are separated by a slit plate. When the vehicle's rate of deceleration changes, the slit plate swings in the vehicle's rear-to-front direction. The slits in the slit plate act to expose the light from the LEDs to the phototransistors. This movement of the slit plate switches the phototransistors ON and OFF. The combinations formed by the two phototransistors switching ON and OFF distinguish the rate of deceleration into four levels, which are sent as signals to the ABS ECU.

A *steering angle sensor* (SAS) measures the steering wheel position angle and rate of turn. The SAS is located in a sensor cluster in the steering column. The cluster always has more than one steering position sensor for redundancy and to confirm data. The ECU module must receive two signals to confirm the steering wheel position. These signals are often out of phase with each other. The SAS tells the ABS control module where the driver is steering the vehicle while the body motion sensors tell it how the body is responding.

A *yaw-rate sensor* is a gyroscopic device that measures a vehicle's angular velocity around its vertical axis. The angle between the vehicle's heading and vehicle actual movement direction is called *slip angle*, which is related to the yaw rate.

A *brake pressure sensor* captures the dynamic pressure distribution between a brake pad and the rotor surfaces during actual braking.

1.2.2.2 Valves and Pumps

Auto brakes typically work with hydraulic fluid. The brake's master cylinder supplies fluid pressure when the pedal is applied. In a standard ABS system, the HCU houses electrically operated hydraulic control *solenoid valves* that control the brake pressure to specific wheel brake circuits. A solenoid valve is a plunger valve that is opened and closed electrically. When power is applied to the solenoid, a magnetic coil is energized, which moves the plunger.

There are multiple *hydraulic circuits* within the HCU, and each hydraulic circuit controls a pair of solenoid valves: an *isolation valve* and a *dump valve*. The valves have three operation modes: *apply*, *hold*, and *release*. In the apply mode, both valves are open and allow the brake fluid to freely flow through the

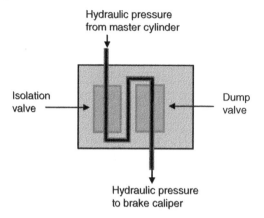

Hydraulic pressure
from master cylinder

Isolation
valve

Dump
valve

Hydraulic pressure
to brake caliper

Figure 1.3 Valves operate in the apply mode.

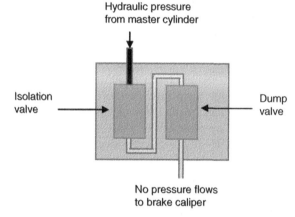

Hydraulic pressure
from master cylinder

Isolation
valve

Dump
valve

No pressure flows
to brake caliper

Figure 1.4 Valves operate in the hold mode.

HCU control circuit to the specific brake circuit, as illustrated in Figure 1.3. In this mode, the driver is in full control of the brakes through the master cylinder.

In the hold mode, both valves are in the closed position that isolates the master cylinder from the brake circuit. This prevents the brake pressure from increasing any further should the driver push the brake pedal harder. The brake pressure to the wheel is held at that level until the solenoid valves are commanded to change its position. This is illustrated in Figure 1.4.

In the release mode, the isolation solenoid valve is closed, but the dump valve is open to release some of the pressure from the brake, allowing the wheel to start rolling again. The dump valve opens a passage back to the accumulator where the brake fluid is stored until it can be returned by an electric *pump* to the master-cylinder reservoir. This is illustrated in Figure 1.5.

The pump is the heart of the ABS. Antilock brakes wouldn't exist without the hydraulic ABS pump. At the detection of wheel slip under heavy breaking,

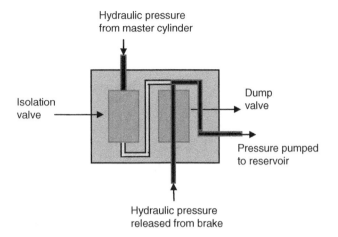

Hydraulic pressure
from master cylinder

Isolation
valve

Dump
valve

Pressure pumped
to reservoir

Hydraulic pressure
released from brake

Figure 1.5 Valves operate in the release mode.

the pump in the HCU sends the brake fluid back to the master cylinder, pushing one or both pistons rearward in the bore. The controller modulates the pump's status in order to provide the desired amount of pressure and reduce slipping.

All valves are open during normal braking. When a wheel locks up, the braking pressure supplied to it should be reduced until it returns to spin. The ABS hydraulic unit works by closing the solenoid valve leading to the wheel that is locking up, thereby reducing the brake force the wheel receives. This way, the wheel's deceleration rate slows down to a safe level. Once that level is achieved, the solenoid opens again to perform its normal function.

1.2.2.3 Electrical Control Unit

The ECU is the brain of the ABS. It is a computer in the car. It watches all the sensors connected to it and controls the valves and pumps, as shown in Figure 1.6. Simply put, if the ABS sensors placed at each wheel detect a lockup, ABS intervenes within milliseconds by modulating the braking pressure at each individual wheel. In this way, the ABS prevents the wheels from locking up during braking, thus ensuring steerability and stability combined with the shortest possible braking distance.

The ECU periodically polls the sensor readings all the time and determines whether any unusual deceleration in the wheels occurs. Normally, it will take a car 5 seconds to stop from 60 mph under ideal conditions, but when there is a wheel lockup, the car could stop spinning in less than 1 second. Therefore, a rapid deceleration in the wheels is a strong indication that a lockup is occurring. When the ECU detects a rapid deceleration, it sends a control signal to the HCU to reduce the pressure to the brakes. When it senses the wheels accelerate, then it increases the pressure until it senses the deceleration

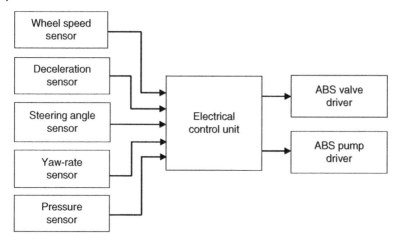

Figure 1.6 Electrical control unit.

again. The deceleration–acceleration repetition occurs very quickly, with the rapid opening and closing of the valves, until the tires slow down at the same rate as the car. Some ABS systems can cycle up to 16 times per second.

1.2.3 ABS Control

ABS brake controllers pose unique challenges to the designers. The main difficulties in the design of any ABS controller arise from the strong nonlinearity and uncertainty of the system to be controlled. First of all, the interaction between the tire and the road surface is very complex and hardly understood. Existing friction models are mostly experimental-based approximations of highly nonlinear phenomena. The dynamics of the whole vehicle is also nonlinear, and it even varies over time. In addition, ABS actuators are discrete, and control precision must be achieved with only three types of control commands: build pressure, hold pressure, or reduce pressure (recall the three operation modes of solenoid valves).

Many different control methods have been developed for ABS. Research on improved control methods is still continuing. The method applied in early systems is *threshold control,* which is as simple as bang–bang control. It uses wheel acceleration and wheel slip as controlled variables. Once the calculated wheel deceleration or wheel slip is over one of the threshold values, the brake pressure is commanded to increase, hold constant, or decrease. Since the brake pressure is cyclically changed based solely on the binary states of the input variables, wheel speed oscillations over time are less controllable.

A class of more advanced control methods uses a cascade closed-loop control structure shown in Figure 1.7. The outer loop, which includes the

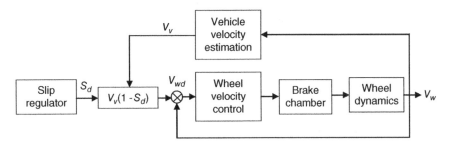

Figure 1.7 A Cascade control structure for ABS.

vehicle velocity estimation (V_v) and desired slip calculation, provides the command signal (V_{wd}) for the inner wheel velocity loop. For inner-loop control, several mechanisms have been proposed. The PID controller, for example, is one of such mechanisms. PID stands for proportional–integral–derivative. PID controllers adopt feedback control law and are widely used in industrial control systems. A PID controller continuously calculates an error value as the difference between a desired set point and a measured process variable. It is an integration of three control policies. With the proportional control (P), the controller output is proportional to the error, which in our case is V_{wd}. With the integral control (I), the controller output is proportional to the amount of time the error is present, which eventually leads to a zero error. With the derivative control (D), the controller output is proportional to the rate of change of the error, which reduces the response time. It has been shown that even for complex and changing surface types, good results can be attained with the conventional PID control algorithms. In the recent years, the well-known features of the conventional PID have been combined with the robustness, self-tuning, or adaptability to nonlinear models of other control methods, which highly enhanced the ABS performance.

No matter what control law is adopted, the feedback control loop can be implemented as an infinite periodic loop:

```
set timer to interrupt periodically with period T;
DO FOREVER
  wait for interrupt;
  read sensor data;
  compute control value u;
  output u;
ENDDO;
```

Here, the period T is a constant in most applications. It is an important engineering parameter. If T is too large, then the control variables will not get adjusted quickly; if it is small, then it will result in excessive computation.

1.3 Real-Time Embedded System Characteristics

The ABS example should have given us some idea about what a real-time embedded system looks like and how it interacts with the larger system it resides in to fulfill the specified function. This section discusses the characteristics of general real-time embedded systems.

1.3.1 System Structure

A real-time embedded system interacts with its environment continuously and timely. To retrieve data from its environment – the target that it controls or monitors, the system must have sensors in place. For example, the ABS has several types of sensors, including wheel speed sensors, deceleration sensors, and brake pressure sensors. In the real world, on the other hand, most of the data is characterized by analog signals. In order to manipulate the data using a microprocessor, the analog data needs to be converted to digital signals, so that the microprocessor will be able to read, understand, and manipulate the data. Therefore, an analog-to-digit converter (ADC) is needed in between a sensor and a microprocessor.

The brain of an embedded system is a controller, which is an embedded computer composed of one or more microprocessors, memory, some peripherals, and a real-time software application. The software is usually composed of a set of real-time tasks that run concurrently, may or may not be with the support of a real-time operating system, depending on the complexity of the embedded system.

The controller acts upon the target system through *actuators*. An actuator can be hydraulic, electric, thermal, magnetic, or mechanic. In the case of ABS, the actuator is the HCU that contains valves and pumps. The output that the microprocessor delivers is a digit signal, while the actuator is a physical device and can only act on analog input. Therefore, a digit-to-analog conversion (DAC) needs to be performed in order to apply the microprocessor output to the actuator. Figure 1.8 shows the relations among all these system components.

1.3.2 Real-Time Response

A real-time system or application has to finish certain tasks within specified time boundaries. This is the character that distinguishes a real-time system from a non-real-time system. The ABS is a typical real-time system. When the sensors detect a dramatic deceleration of wheels, the system must act quickly to prevent the wheels from being locked up; otherwise, a disaster may occur. Moreover, the control law computing is also real-time: a cycle of a sensor data processing and control value computing must be finished before the next cycle starts; otherwise, the data to be processed will pile up. If a missile guidance system fails to make timely corrections to its attitude, it can hit the wrong

Figure 1.8 Structure of real-time embedded systems.

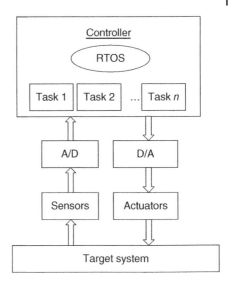

target. If a GPS satellite doesn't keep a highly precise measure of time, position calculations based on its signal will simply be wrong.

Deadlines of real-time tasks are typically derived from the required responsiveness of the sensors, actuators, and the dynamics of the target that the embedded system controls. Real-time systems are expected to execute all tasks by their deadlines. However, "real-time" does not mean "real fast" or "the faster, the better." Take a cardiac pacemaker as an example. Obviously, if it fails to induce current through the heart muscle at the right time, the patient's heart can go into fibrillation. However, if it induces current faster than normal heart rhythm, it will cause problem as well.

1.3.3 Highly Constrained Environments

Real-time embedded systems are often run in highly resource-constrained environments, which make the system design and performance optimization quite challenging. Although some embedded systems, such as air-traffic control systems and wireless mobile communication systems, run with very powerful processors, a lot of them are equipped with 8-bit processors only. Examples are the systems embedded in dishwashers, microwaves, coffee makers, and digital watches. Most embedded systems are constrained in terms of processor speed, memory capacity, and user interface. Many embedded systems operate in an uncontrolled harsh environment. They have to survive excessive heat, moisture, vibration, shock, or even corrosion. The ABS and automotive embedded systems that control ignition, combustion, and suspension are such examples. Therefore, embedded systems must be optimized in terms of size, weight, reliability, performance, cost, and power consumption to fit into the computing environment and perform their tasks. Thus, embedded systems typically require far more optimization than standard desktop applications.

1.3.4 Concurrency

Concurrency refers to a property of systems in which several computations are executing simultaneously and potentially interacting with each other. Embedded systems by design are closely interacting with their physical environment. We have demonstrated this feature through the ABS analysis. Physical environment is by its nature concurrent – multiple processes occur at the same time. For example, the following events in the ABS can occur at the same time:

- Wheel speed sensor event
- Deceleration sensor event
- Brake pedal event
- Solenoid valve movement
- Pump operation

Almost all of these events have strict constraints on the response time. All deadlines should be met.

Because of the existence of multiple control processes and each process may have its own control rate, many real-time embedded systems are *multirate* systems. For example, a real-time surveillance system needs to process both audio and video inputs, but they are processed at different rates.

1.3.5 Predictability

A real-time system must behave in a way that can be predicted in terms of all timing requirements. For instance, it must be mathematically predictable if a specific task can be completed before a given deadline. Factors that go into this calculation are system workload, the power of processors, run-time operating system support, process and thread priorities, scheduling algorithm, communication infrastructure, and so on. A real-time system such as an ABS, or an airplane's flight-control system, must always be 100% predictable, or human lives are at stake.

Many real-time embedded systems contain heterogeneous computing resources, memories, bus systems, and operating systems and involve distributed computing via global communication systems. Latency and jitter in events are inevitable in these systems. Therefore, related constraints should be specified and enforced. Otherwise, these systems can become unpredictable.

A term related to predictability is determinism. Determinism represents the ability to ensure the execution of an application without concern that outside factors, such as unforeseen events, will upset the execution in unpredictable ways. In other words, the application will behave as intended in terms of functionality, performance, and response time, all of the time without question.

1.3.6 Safety and Reliability

Some real-time embedded systems are safety-critical and must have high reliability. Examples are cardiac pacemakers and flight control systems. The term *safety* means "freedom from accidents or losses" and is usually concerned with safety in the absence of faults as well as in the presence of single-point faults. Reliability, on the other hand, refers to the ability of a system or component to perform its required functions under stated conditions for a specified time. It is defined as a stochastic measure of the percentage of the time the system delivers services. Embedded systems often reside in machines that are expected to run continuously for years without errors. Some systems, such as space systems and undersea cables, are even inaccessible for repair. Therefore, embedded system hardware and software are usually developed and tested more carefully than those for general-purpose computing systems.

Reliability is often measured in failures per million operating hours. For example, the requirement for a typical automotive microcontroller is 0.12 failures per million operating hours. The measurement is 37.3 for an automotive oil pump. Failures could be caused by mechanical "wear-out," software defects, or accumulated run-time faults.

1.4 Hard and Soft Real-Time Embedded Systems

There are *hard* real-time systems and *soft* real-time systems. A hard real-time system is a system in which most timing constraints are hard. A soft real-time system is a system in which most timing constraints are soft.

A hard real-time constraint is a constraint that a system *must* meet. If the deadline is missed, it will either cause the system failure or result in a zero usefulness of the delivered service. On the other hand, a soft constraint is a constraint that a system should meet, but when the deadline is occasionally missed, it won't cause any disastrous result, and the delivered service is still useful to a certain extent.

Normally, a hard constraint is expressed deterministically. For example, we may have the following constraints for the ABS:

- The wheel speed sensors must be polled every 15 milliseconds.
- Each cycle, the control law computation for wheel speed must be finished in 20 milliseconds.
- Each cycle, the wheel speed prediction must be completed in 10 milliseconds.

These constraints are hard because the sensor data, control value, and wheel speed predicted value are all critical to the correct functioning of the ABS. It is

Figure 1.9 Value of hard and soft real-time tasks when deadline is missed.

also because these events are periodical. If the deadline of one cycle is missed, the next cycle starts immediately, and thus, the late result becomes useless.

Soft constraints are often expressed statistically. For example, we may have the following constraints for an automated teller machine (ATM):

- After a credit card or debit card is inserted, the probability that the ATM prompts the user to enter a passcode within 1 second should be no less than 95%.
- After it receives a positive response from the bank that issued the card, the ATM should dispense the specified amount of cash within 3 seconds at a chance of no less than 90%.

The deadlines with these two constraints are soft, because a few misses of the deadlines do no cause serious harm; only the degree of customers' satisfaction of using the system is negatively affected.

Figure 1.9 illustrates the value function of a real-time task in terms of response time. If the task has a hard deadline, then its value drops to zero when the deadline is missed. If the task has a soft deadline, then its value decreases when the deadline is missed, but not to zero right way.

Many hard real-time systems also have soft constraints and vice versa. When a timing constraint is specified as hard, then a rigorous validation is required.

Exercises

1   Give an example of real-time database application. Is it a hard or a soft real-time system? Give your arguments.

2   The car engine management system (EMS) is a real-time embedded system. Read related online materials, and find out major hardware components of the system and how they interact with each other to ensure the best engine performance.

3   Give an example of real-time embedded systems in which an earlier response than expected is as bad as a late response.

4   Give an example of a real-time embedded system that has both hard and soft real-time constraints.

Suggestions for Reading

Shin and Ramanathan [1] introduces the basic concepts and identifies the key issues in the design of real-time systems. Axer *et al.* [2] summarizes the current state of the art in research concerning how to build timing-predictable embedded systems. A survey of ABS control laws is presented in Ref. [3].

References

1 Shin, K. and Ramanathan, P. (1994) Real-time computing: a new discipline of computer science and engineering. *Proceedings of the IEEE*, **82** (1), 6–25.
2 Axer, P., Ernst, R., Falk, H. *et al.* (2012) *Building Timing Predictable Embedded Systems*, ACM Transactions on Embedded Computing Systems.
3 Aly, A., Zeidan, E., Hamed, A., and Salem, F. (2011) An antilock-braking systems (ABS) control: a technical review. *Intelligent Control and Automation*, **2**, 186–195.

2

Hardware Components

This chapter introduces real-time embedded system hardware components. Because real-time embedded systems range from small ones such as those in coffee makers and digital watches to big and sophisticated ones such as railroad control systems and mobile communication switches, there is a big difference in the set of hardware components used. Figure 2.1 shows a set of typical embedded system hardware units.

2.1 Processors

The processors used in embedded systems vary with the need of computation power of individual embedded application. They fall into two general categories, however. One is general-purpose microprocessors, and the other is special-purpose processors. Microcontrollers and application-specific integrated circuits (ASICs) are the most popular special-purpose processors.

2.1.1 Microprocessors

Many real-time embedded systems use general-purpose microprocessors. A microprocessor is a computer processor on an integrated circuit. It contains all, or most of, the central processing unit (CPU) functions. Figure 2.2 shows a set of elements that are necessary for microprocessors to perform operations.

A microprocessor is designed to perform arithmetic and logic operations that make use of small storage called *registers*. It has a *control unit* that is responsible for directing the processor to carry out stored program instructions. It communicates with both the *arithmetic logic unit* (ALU) and memory. All the instructions that are fetched from memory are stored in the *instruction register* as binary values. The *instruction decoder* reads that values and tells the ALU which computational circuits to energize in order to perform the function. The ALU performs integer arithmetic and bitwise logic operations. These operations are the result of instructions that are part of the microprocessor design.

Real-Time Embedded Systems, First Edition. Jiacun Wang.
© 2017 John Wiley & Sons, Inc. Published 2017 by John Wiley & Sons, Inc.

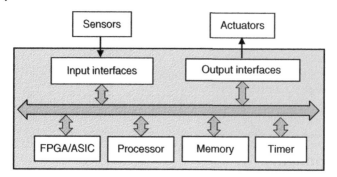

Figure 2.1 Real-time embedded system hardware components.

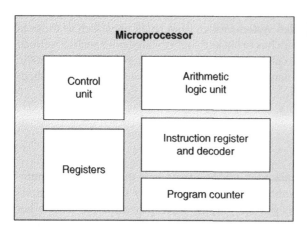

Figure 2.2 Microprocessor elements.

Program counter stores the address of the next instruction to be executed. Microprocessors operate on numbers and symbols represented in the binary numeral system.

The first commercial microprocessor Intel 4004 was invented by Federico Faggin and his coworkers in the early 1970s. Intel 4004 is a 4-bit CPU released by Intel Corporation. Before that, small computers had been built using racks of circuit boards with many medium- and small-scale integrated circuits (ICs). Microprocessors combined them into one or a few large-scale ICs. This approach of CPU implementation overtook all other central processing unit implementation methods quickly. Most modern microprocessors are either 32-bit or 64-bit, although 128-bit microprocessors are also available. Examples of general-purpose microprocessors include Intel 80x86, SPARC, and Motorola 68HCxxx.

Microprocessors find applications where tasks are unspecific. For example, they can be used for developing software, games, and websites, editing photos, or creating documents. In such cases, the relationship between the input and output is not defined. They need a large number of resources, such as RAM, ROM, and I/O ports. The embedded software can be tailored for specific tasks that are designed for an embedded system.

2.1.2 Microcontrollers

Compared to a general-purpose microprocessor, a *microcontroller* is a self-contained system with peripherals, memory, and a processor that is designed to perform specific tasks. Microcontrollers are used in systems where the relationship between the input and output is usually clearly defined. Examples are a computer mouse, washing machine, digital camera, microwave, car, cell phone, and digital watch. Since the applications are very specific, they have little demand on resources such as RAM, ROM, I/O ports, and hence, they can be embedded on a single chip with the processor. This in turn reduces the size and the cost. A microcontroller is cheap to replace, while microprocessors are 10 times more expensive. Moreover, microcontrollers are generally built using a technology known as complementary metal–oxide–semiconductor (CMOS). This technology is a competent fabrication system that uses less power and is more immune to power spikes compared to other techniques.

Examples of commonly used 16-bit microcontrollers for medium-scale embedded systems are PIC24 series, Z16F series, and IA188 series. The most common sizes for RAM are 1.5, 2, 4, 8, 16, and 32 kB.

2.1.3 Application-Specific Integrated Circuits (ASICs)

An ASIC is a highly specialized device and constructed for one specific-purpose application only. It is used as a replacement to general-purpose logic circuitry. It integrates several functions into a single chip and thus reduces the number of overall circuits needed. ASICs are very expensive to manufacture, and once it is made, there is no way to modify or improve, as the metal interconnect mask set and its development are the most expensive and of fixed cost. If you want to alter the instruction set, or do something similar, you have to modify the actual silicon IC layout. The lack of programmability and high cost make ASICs not suitable for use in the prototyping stage of system design cycle.

ASICs are widely used in communication, medical, network, and multimedia systems, such as cellular phones, network routers, and game consoles. Most SoC (Systems-on-a-Chip) chips are also ASICs. A microcontroller can be viewed as a type of ASIC that executes a program and can do generic things as a result. For a given application, ASIC solutions are normally more effective than the solutions based on the software running on microprocessors.

2.1.4 Field-Programmable Gate Arrays (FPGAs)

An FPGA is a *programmable* ASIC. It contains a regular grid of logic cells that can be rapidly reconfigured, which facilitates fast prototyping of embedded systems. FPGAs are commonly used during system design. They are usually replaced in the final product with custom circuitry, such as ASIC chips, due to higher performance and lower cost. When reconfigurability is an essential part of the functionality of a real-time embedded system, FPGAs do appear in the final product.

The first commercial FPGA was developed by Xilinx in 1985. Modern FPGAs are fabricated using the most advanced technology and enable implementation of very high performance systems. For example, the latest Xilinx Virtex Ultra-Scale is built on a 20-nm technology, introduced in May, 2014. The UltraScale uses a 3D or stacked architecture that contains up to 4.4 million logic cells.

An FPGA can be used to solve any problem that is computable. Specific applications of FPGAs include digital signal processing, software-defined radio, ASIC prototyping, medical imaging, computer vision, speech recognition, cryptography, bioinformatics, computer hardware emulation, radio astronomy, metal detection, and a growing range of other areas. In 2013, the market of FPGAs was $5.4 billion and is estimated to reach $9.8 billion by 2020.

2.1.5 Digital Signal Processors (DSPs)

DSPs are designed for high-data-rate computations. DSPs implement algorithms in hardware and offer high performance in repetitive and numerically intensive tasks. DSPs are two to three times faster than the general-purpose microprocessors in signal processing applications, including audio, video, and communication applications.

Disadvantages include generally high cost. A recent study also indicates that many commercially available DSPs lack adequate compiler support.

2.1.6 Application-Specific Instruction Set Processors (ASIPs)

ASIPs are an emerging design paradigm that offers an intermediary solution between ASICs and programmable processors. Typically, ASIPs consist of custom integrated circuitry that is integrated with an instruction set tailored to benefit a specific application. This specialization of the core provides a trade-off between the flexibility of a general-purpose processor and the performance of an ASIC.

Advantages of ASIPs include high performance and increased design flexibility because later design changes can be accommodated by updating the application software running on the ASIP.

2.1.7 Multicore Processors

Because of the fact that the processor clock speed is closely linked to the number of transistors that can fit on a chip, when transistor shrinking technology began to slow down, improvement in increased processor speed also began to slow down. There is also a power wall issue. That is, as processors become more capable (denser transistors on a chip), their energy consumption and heat production increase rapidly as well. As a result, multicore processing has become a growing industry trend. Most current systems are multicore. A multicore processor is an integrated circuit to which two or more processors have been attached for enhanced performance, reduced power consumption, and more efficient simultaneous processing of multiple tasks.

Multicore is a shared memory multiprocessor: all cores share the same memory. However, each core typically has its own private cache memory. Threads of different cores run in parallel. Within each core, threads are time-sliced.

2.1.8 Von Neumann Architecture and Harvard Architecture

There are two fundamental computer architectures, namely von Neumann architecture and Harvard architecture. Both architectures use stored program mechanism, which keeps program instructions and data in read–write, random-access memory (RAM). The difference is that the von Neumann architecture uses common memory to store both data and instructions, and thus, an instruction fetch and a data operation cannot occur at the same time because they share a common bus. The Harvard architecture, on the other hand, separates the storage of instructions from data. This way, the CPU can both read an instruction and perform a data memory access at the same time. The two architectures are illustrated in Figure 2.3.

The von Neumann architecture shares single common bus for instruction and data, which results in a low performance as compared to the Harvard architecture. It is often called the *von Neumann bottleneck*. Several approaches to overcoming the von Neumann bottleneck have been developed. One is to use cache memory. We will discuss it later. Moreover, accidental corruption of program memory may occur with the von Neumann architecture, because data memory and program are stored physically in the same chip. However, since data memory and program memory are stored physically in different locations in

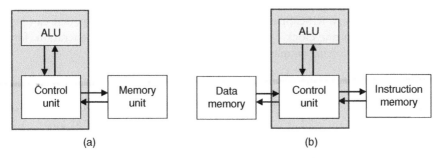

Figure 2.3 Two fundamental computer architectures. (a) von Neumann architecture and (b) Harvard architecture.

the Harvard architecture, no chances exist for accidental corruption of program memory.

Most DSPs use Harvard architecture for streaming data to achieve greater and more predictable memory bandwidth.

2.1.9 Complex Instruction Set Computing and Reduced Instruction Set Computing

An *instruction set* is a group of instructions that can be input to the processor. These instructions direct the processor in terms of data manipulation. An instruction typically includes an opcode that specifies the operation to perform, such as add contents of memory to register, and zero or more operand specifiers, which may specify registers, memory locations, or literal data. An *instruction set architecture* serves as an interface to allow easy communication between the programmer and the hardware. It prepares the processor to respond to all the user commands.

There are two prevalent instruction set architectures: complex instruction set computing (CISC) and reduced instruction set computing (RISC). CISC processors run *complex instructions* where a single instruction may execute several low-level operations. The primary goal of CISC architecture is to complete a task in as few lines of assembly instructions as possible. Let us say we want to multiply two numbers. A CISC processor would come prepared with a specific instruction, say MULT. When executed, this instruction will load the two numbers from the main memory into two separated registers, multiply them in the execution unit, and then store the product in an appropriate register or back to memory. Thus, the entire task of multiplying two numbers is completed with a single instruction such as

```
MULT A, B
```

MULT is a complex instruction. It operates directly on the computer's memory banks and does not require the programmer to explicitly call any loading or storing functions. Executing a complex instruction may need multiple clock cycles.

On the contrary, RISC processors only use simple instructions that can be executed within one clock cycle. To perform a multiplication operation, for example, the following simple instructions are required:

```
LOAD R1, A      ; load A into register R1
LOAD R2, B      ; load B into register R2
PROD R1, R2     ; multiply A and B, product saved in R1
STOR R1, A      ; store A*B into a memory location
```

Examples of CISC processors are Intel x86 and SHARC. Examples of RISC processors are ARM 7 and ARM 9.

2.2 Memory and Cache

Memory is one of the basic components of embedded systems. The fundamental building block of memory is *memory cell*. The memory cell is an electronic circuit that stores one bit of binary information, and it must be set to store a logic 1 (high voltage level) and reset to store a logic 0 (low voltage level). Its value is maintained until it is changed by the set/reset process. Memory is used to store both programs and data.

One can envision memory as a matrix of bits, where the length of each row is the size of the addressable unit of the memory. The total number of rows represents the memory's capacity. Each row is implemented by a register and has a unique address. Memory addresses typically start from 0 and grow upward. Normally, memory is *byte-addressable*, meaning each register has 8 bits. Some machines can process 32 bits a time, and the memory is implemented with 32-bit registers. We say the memory is 32-bit word-addressable.

2.2.1 Read-Only Memory (ROM)

ROM is used to store programs. While a program is running, the data in the program memory won't change. ROM is a type of nonvolatile memory, and the stored program won't be lost when the ROM is powered off. ROM is hardwired and cannot be changed after manufacture.

PROM (Programmable Read-Only Memory) is similar to ROM except that it is programmable. We can buy a blank chip and have a PROM programmer program it to meet our requirements. But, once we program it, we can never change it.

EPROM (Erasable Programmable Read-Only Memory) is also a nonvolatile memory. What makes it distinct from ROM or PROM is that once programmed, an EPROM can be erased by exposing it to strong ultraviolet light source (such as from a mercury-vapor light), and then a new program can be written into it.

EEPROM (Electrically Erasable Programmable Read-Only Memory) is used similarly to the hard drive in a personal computer, to store settings that might change occasionally, which need to be remembered next time the computer starts up. Essentially, it can be written, erased, rewritten electronically. No special treatment is required to erase the data.

Flash is the latest ROM and the most popular technology used in today's embedded design. Flash memory is an electronic (solid-state) nonvolatile storage medium that can be electrically erased and reprogrammed. Flash memory is technically a type of EEPROM. Flash memory earned its name because of its high speed (similarly to the flash of a camera) in erasing all the data from a semiconductor chip.

2.2.2   Random-Access Memory (RAM)

RAM is the simplest and most common form of data storage. RAM allows data items to be read or written in almost the same amount of time irrespective of the physical location of the data inside the memory. Unlike ROM, RAM is volatile, which means a power off will erase all the data in the RAM. There are two types of RAM that are widely used, namely *static* RAM (SRAM) and *dynamic* RAM (DRAM). They use different technologies for data storage. SRAM uses six transistors per memory cell, whereas DRAM uses only one transistor per memory cell. Therefore, SRAM is more expensive to produce. However, SRAM cell is a type of flip-flop circuit, usually implemented using field-effect transistors that have high input impedance. It requires very low power when not being accessed.

DRAM uses capacitive storage. The capacitor holds a high or low charge (1 or 0, respectively), and the transistor acts as a switch that lets the control circuitry on the chip read the capacitor's state of charge or change it. Since the capacitor can lose charge, DRAM needs to be refreshed periodically. This makes DRAM more complex and power-consuming. However, as this form of memory is less expensive to produce compared to SRAM, it is the predominant form of memory used in embedded systems.

SDRAM (*synchronous* DRAM) is a type of DRAM that is synchronized with the system bus. It is a generic name for various kinds of DRAM that are synchronized with the clock speed that the microprocessor is optimized for. SDRAM can accept one command and transfer one word of data per clock cycle. It is capable of running at 133 MHz, a typical clock frequency.

2.2.3   Cache Memory

Processor speed has significantly increased in recent years. Memory improvements, on the other hand, have mostly been in terms of density – the ability to store more data in less space – rather than transfer rates. When a fast processor works with a slow memory, the overall speed is low, because no matter how fast the processor can work, its actual speed is limited to the rate of data transfer from the memory. Therefore, a faster processor just means that it spends more time being idle.

Cache memory technology is developed to overcome this problem. Cache memory is a type of RAM that a microprocessor can access significantly faster than it can access regular RAM. This memory is typically integrated directly with the processor chip or placed on a separate chip that has a separate bus interconnected with the processor. Cache memory is used to store program instructions that are frequently re-referenced by software during operation, whereas less frequently used data is stored in a big and low-speed memory device. When the processor processes data, it looks first in the cache memory; if it finds the instructions there, it does not have to do a more time-consuming

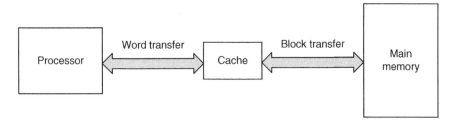

Figure 2.4 Cache memory.

reading of data from regular memory. If not, a block of main memory, consisting of some fixed number of words, is read into the cache, and then the word is delivered to the processor. The other words in the block are likely to be referenced in the near future due to the phenomenon of locality of reference. Such a design increases the overall speed of the software execution. The idea is illustrated in Figure 2.4.

Some CPUs have multilevel cache memory. They are referred to as L1 for level one, L2 for level two, L3 for level three, and so on. A CPU directly works with L1 cache. The goal of the L1 cache is to line up with the CPU so that it has data to work on every processor cycle at the best. It is a small, high-speed cache incorporated right onto the processor's chip. The L1 cache typically ranges in size from 8 to 64 KB and uses the high-speed SRAM. The L2 cache is expected to feed data to the L1 cache every few processor cycles, while the L3 cache can feed data to the L2 cache at a further slower rate. The three levels exchange data based on CPU requirements. Figure 2.5 shows a dual-core processor with two levels of cache memory. Each core has its private L1 cache, but they share the L2 cache and, of course, the main memory.

Figure 2.5 A dual-core processor with two levels of cache memory.

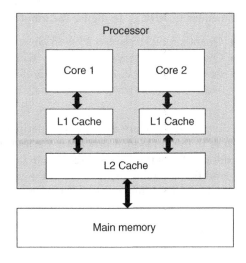

2.3 I/O Interfaces

Embedded processors communicate with the outside world through I/O interfaces. An I/O interface is an electronic device, exposed to designers as pins of the chip, which has one side connecting to the processor and the other side connecting to input/output devices. For example, a key pad in a microwave is an electronic device for users to interact with the microwave. There is a special circuit between the key pad and the embedded microcontroller that serves as the interface. When a key is pressed by a user, the circuit picks up the event, converts it into a unique recognizable binary number, and presents it to the processor's I/O port. The processor reads the number from the port and processes it based on what it means.

In case of analog I/O, as seen in the example of ABS, an A/D converter (ADC) is needed to encode analog inputs (normally voltage) into a digital word (normally 8 bits or 16 bits), and a D/A converter (DAC) is needed to decode digital outputs. Let V_{RefL} and V_{RefH} be the lower bound and upper bound, respectively, of voltage that an ADC can encode. The ADC encodes all possible inputs in $[V_{RefL}, V_{RefH}]$ into a range of values from 00 to FF (8 bits) or from 0000 to FFFF (16 bits). The number of bits in the digit word determines the resolution of the conversion, which represents the magnitude of the quantization error. For example, an ADC with a resolution of 8 bits can encode an analog input to one in 256 different levels.

The ADC resolution can also be expressed in volts. The minimum change in voltage required to guarantee a change in the output code level is called the *least significant bit* (LSB) voltage. The resolution Q of the ADC is equal to the LSB voltage. The maximum quantization error is half of the LSB voltage.

Example 2.1 *ADC Resolution and Quantization Error*
Consider an ADC of 16 digits. The input voltage ranges from 0 to 5 V.

- Quantization levels: $L = 2^{16} = 65536$
- Quantization intervals: $I = N - 1 = 65535$
- Resolution: $Q = (V_{refH} - V_{refL})/I = 5/65535 = 0.000076$ V
- Maximum quantization error: $E = Q/2 = 0.000038$ V

Typically, there are multiple I/O peripherals connected through I/O interfaces to the embedded processor. Each peripheral is identified with a unique address. When the process executes an I/O related instruction, it issues a command that contains the address of the target peripheral. Therefore, each I/O interface must interpret the address lines to determine if the command is for itself.

In a design where the processor, main memory, and I/O peripherals share a common bus, there are two complementary ways the I/O is mapped:

port-mapped I/O (also called *isolated* I/O) and *memory-mapped* I/O. Port-mapped I/O uses a separate address space from main memory, accomplished by an extra I/O pin on the CPU's physical interface, or a dedicated bus to I/O. I/O devices are accessed via a dedicated set of microprocessor instructions. Because the address space for I/O is isolated from that for main memory, this is sometimes referred to as isolated I/O.

Memory-mapped I/O means mapping the I/O peripherals' memory into the main memory map. That is, there will be addresses in the processor's memory that won't actually correspond to RAM, but to memory of peripherals. Compared to port-mapped I/O, memory-mapped I/O is simple in design, because port-mapped I/O requires either additional pins in the processor or an entire separate bus, while memory-mapped I/O does not have this extra complexity.

Memory-mapped I/O is also more efficient than port-mapped I/O. Port-mapped I/O instructions are very limited in capability, often provided only for simple load-and-store operations between CPU registers and I/O ports. Because of that, to add a constant to a port-mapped device register would require three instructions: read the port to a CPU register, add the constant to the CPU register, and write the result back to the port. However, because regular memory instructions are used to address the devices in memory mapper I/O, all of the CPU's addressing modes are available for I/O as well as memory, and instructions that perform an ALU operation directly on a memory operand can be used with I/O device registers as well.

2.4 Sensors and Actuators

A sensor is an input device of embedded systems. It is a transducer that converts energy from one form to another for measurement or control purpose. For example, an ultrasonic sensor converts ultrasound waves to electrical signals, an accelerometer converts acceleration to voltage, and a camera is a sensor that converts photon energy to electrical charge that represents the photon flux for each picture element in an array. There are many types of sensors. Displacement sensors, pressure sensors, humidity sensors, acceleration sensors, gyro sensors, temperature sensors, and light sensors are among the most widely used ones. A good sensor must be sensitive to the measured property, but does not interfere with it. Recall the ABS wheel speed sensor that is introduced in Chapter 1. The input of the sensor is the rotation of the sensor rotor, and the output is AC voltage. The voltage magnitude and frequency are proportional to the rotation rate.

Sensors can be designed for virtually every physical and chemical quantity, including weight, velocity, acceleration, electrical current, voltage, temperatures, and chemical compounds. Many physical effects are used for constructing the sensors. For example, the automotive wheel speed sensors use the *induction effect*, that is, when a magnetic field interacts with an electric circuit, an electromotive force is produced. The automotive airbag sensor is designed based on the effect *piezoelectric effect*; that is, certain materials generate an electric charge in response to the applied mechanical stress.

There are *active sensors* and *passive sensors*. An active sensor requires an external source of energy to operate. Examples are radar, sonar, GPS and X-ray. On the contrary, passive sensors simply detect and respond to some type of input from the physical environment. For example, the wheel speed sensor is a passive sensor. It detects and measures the wheel rotation without the need to send any signal or apply any energy to the wheel. Temperature sensors are another example of passive sensors.

The performance of sensors is mainly characterized by the following parameters:

- Range of the value of the measured stimulus
- Resolution of the measured stimulus
- Sensing frequency
- Accuracy of measurement
- Size
- Operating temperature and environment conditions
- Service life in hours or number of cycles of operation

Of course, cost is also a concern in sensor selection.

An actuator is a transducer that converts electrical energy into some other form of energy, such as motion, heat, light, or sound, to move or control a system. It provides the driving force for a variety of natural and man-made requirements. For example, the ABS hydraulic control unit introduced in Chapter 1 is an actuator that acts on the ECU output to build, hold, or reduce the brake pressure. Traditional actuators include hydraulics, pneumatics, and solenoids. A hydraulic actuator consists of a cylinder or fluid motor that uses hydraulic power to facilitate mechanical operation. A pneumatic actuator converts energy formed by vacuum or compressed air at a high pressure into either linear or rotary motion. A solenoid is a type of electromagnetic actuator that converts an electrical signal into a magnetic field and produces a linear motion. Solenoids are used in the ABS hydraulic control unit to move the valves.

Some newly developed actuators, such as piezoelectric, shape memory alloy, and magnetostrictive devices, are based on shape-changing materials. They are increasingly used in novel applications. For example, piezoelectric actuators are high-speed precision ceramic actuators that convert electrical energy into linear motion with high resolution. These actuators are used in many modern high-tech areas such as microscopy, bionanotechnology, and astronomy/aerospace technology.

Different types of actuators exhibit different characteristics. Nevertheless, the performance of an actuator is primarily characterized by the following parameters:

- Maximum magnitude of force or mechanic it can exert on a system in sustainable cyclic operation
- Speed of operation
- Operating temperature and environment conditions
- Service life in hours or number of cycles of operation

2.5 Timers and Counters

Timing functions are vital in real-time embedded systems. A timer is a specialized type of clock that is used to measure time intervals. A counter counts the number of external events occurring on its external event pin. When the event is clock pulse, a timer and counter are essentially the same. Therefore, in many occasions, these two terms are used interchangeably.

The main component of a timer is a free-running binary counter. The counter increments for each incoming timing pulse. Since it runs freely, it can count the inputs, which could be clock pulses, while the processor is executing the main program. If the input pulses arrive at a fixed rate, then the pulse count accurately measures the time interval. For example, if the rate of the input pulses is 1 MHz and the counter has recorded 1000 pulses, then the elapsed time is 1000 microseconds. When the count overflows, an output signal is asserted. The overflow signal may thereby trigger an interrupt to the processor or set a bit that the processor can read. Figure 2.6 shows a 16-bit counter that takes clock cycles as input.

The input pulses could be different from clock pulses. In that case, a *prescaler* is used to generate pulses. A prescaler is a configurable clock-divider circuit. It takes the basic clock frequency and divides it by some value before feeding it to the counter. With a prescaler, we can let the counter count at a desired rate. For example, if we configure the prescaler to divide the 1 MHz of clock frequency by 8, then the new rate of the timer will be $10^6/8 = 125$ KHz, and thus, a 16-bit timer can record up to $65,535 \times 8 = 524,280$ microseconds before it overflows.

The free-running counter is often connected to a *capture register*, as illustrated in Figure 2.7. A capture register can automatically load the current

Figure 2.6 Structure of a 16-bit counter.

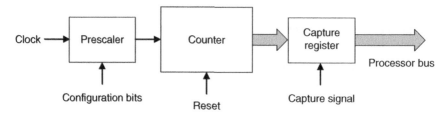

Figure 2.7 A timer with a prescaler and a capture register.

output of the free-running counter upon the occurrence of some event, typically a signal to an input pin, latch the value into a processor-visible register, and then generate an output signal. One use of a timer with the capture requester is to measure the time between the leading edges of two pulses. By reading the value in the capture register and comparing it with a previous reading, the software can determine how many clock cycles elapsed.

Example 2.2 *Timing Events*
Consider a timer that is designed with a prescaler. The prescaler is configured with 3 bits, and the free-running counter has 16 bits. The timer counts the timing pulses from a clock whose frequency is 8 MHz. Suppose that a capture signal from the processor latches a count of 304D in hex. We want to find out how much time had elapsed since the last reset to the free counter.

First, we convert the hex number 304D to the corresponding decimal number, which results in 12,365.

Because the prescaler is configured with 3 bits, it divides the clock frequency by 2^3. Thus, the frequency of the signals that are fed into the free-running counter is $8*10^6/2^3$ Hz or 1 MHz. Therefore, the elapsed time is

$$12365/10^6 = 0.12365 \text{ second} = 12,365 \text{ microseconds.}$$

Exercises

1 Assume a 1 MB memory.
   (a) What are the lowest and highest addresses if it is byte-addressable?
   (b) What are the lowest and highest addresses if it is 32-bit word-addressable?

2 Explain the difference between port-mapped I/O and memory-mapped I/O.

3 Compare the von Neumann architecture with the Harvard architecture.

4 Consider an A/D converter with a full-scale measurement range of −5 to 5 V and a resolution of 16 bits. How many quantization levels are there? What is the maximum quantization error?

5 Consider a timer with a prescaler that is configured with 4 bits, a 16-bit free counter, and a capture register. The input clock frequency is 33 MHz. A capture signal from the processor latches a count of C17E in hex. How much time had elapsed since the last reset to the free counter?

Suggestions for Reading

Instruction set architecture in general and Intel x86 architecture in particular are well discussed in Ref. [1]. The details of SPARC RISC architecture are presented in Ref. [2]. ARM RISC architecture and powerful 32-bit instructions set are introduced in Ref. [3]. References for various microcontroller products and their applications in embedded system design are available. Examples include McKinlay [4] for Intel 8051, Valvano [5] for Texas Instruments MSP432, and Wilmshurst [6] for Microchip PIC products. A good coverage of sensors and actuators can be found in Refs [7] and [8]. If you are interested in building your own embedded system, Catsoulis [9] is a good reference.

References

1 Shanley, T. (2010) *x86 Instruction Set Architecture*, MindShare Press.
2 Paul, R. (1999) *SPARC Architecture, Assembly Language Programming, and C*, 2nd edn, Pearson.
3 Mazidi, M.A. (2016) *ARM Assembly Language Programming & Architecture*, Kindle edn, Micro Digital Ed.
4 McKinlay, M. (2007) *The 8051 Microcontrollers & Embedded Systems*, Pearson.
5 Valvano, J.W. (2015) *Embedded Systems: Introduction to the MSP432 Microcontroller*, CreateSpace Independent Publishing Platform.
6 Wilmshurst, T. (2009) Designing Embedded Systems with PIC Microcontrollers, Principles and Applications, 2nd edn, Newnes.
7 Ida, N. (2013) *Sensors, Actuators, and Their Interfaces: A Multidisciplinary Introduction*, SciTech Publishing.
8 de Silva, C.W. (2015) *Sensors and Actuators: Engineering System Instrumentation*, 2nd edn, CRC Press.
9 Catsoulis, J. (2005) *Designing Embedded Hardware: Create New Computers and Devices*, 2nd edn, O'Reilly Media.

3

Real-Time Operating Systems

The heart of many computerized embedded systems is real-time operating system (RTOS). An RTOS is an operating system that supports the construction of applications that must meet real-time constraints in addition to providing logically correct computation results. It provides mechanisms and services to carry out real-time task scheduling, resource management, and intertask communication. In this chapter, we briefly review the main functions of general-purpose operating systems, and then we discuss the characteristics of RTOS kernels. After that, we introduce some widely used RTOS products.

3.1 Main Functions of General-Purpose Operating Systems

An operating system (OS) is the software that sits between the hardware of a computer and software applications running on the computer. An OS is a resource allocator and manager. It manages the computer hardware resources and hides the details of how the hardware operates to make the computer system more convenient to use. The main hardware resources in a computer are processor, memory, I/O controllers, disks, and other devices such as terminals and networks.

An OS is a policy enforcer. It defines the rules of engagement between the applications and resources and controls the execution of applications to prevent errors and improper use of the computer.

An OS is composed of multiple software components, and the core components in the OS form its *kernel*. The kernel provides the most basic level of control over all of the computer's hardware devices. The kernel of an OS always runs in *system mode*, while other parts and all applications run in *user mode*. Kernel functions are implemented with protection mechanisms such that they could not be covertly changed through the actions of software running in user space.

Real-Time Embedded Systems, First Edition. Jiacun Wang.

The OS also provides an application programming interface (API), which defines the rules and interfaces that enable applications to use OS features and communicate with the hardware and other software applications. User processes can request kernel services through *system call* or by *message passing*. With the system call approach, the user process applies traps to the OS routine that determines which function is to be invoked, switches the processor to system mode, calls the function as a procedure, and then switches the processor back to user mode when the function completes and returns control to the user process. In the message passing approach, the user process constructs a message that describes the desired service and then uses a *send* function to pass it to an OS process. The *send* function checks the desired service specified in the message, changes the processor mode to system mode, and delivers it to the process that implements the service. Meanwhile, the user process waits for the result of the service request with a message *receive* operation. When the service is completed, the OS process sends a message back to the user process.

In the rest of this section, we briefly introduce some of the main functions of a typical general-purpose OS.

3.1.1 Process Management

A *process* is an instance of a program in execution. It is a unit of work within the system. A program is a *passive* entity, while a process is an *active entity*. A process needs resources, such as CPU, memory, I/O devices, and files, to accomplish its task. Executing an application program involves the creation of a process by the OS kernel that assigns memory space and other resources, establishes a priority for the process in multitasking systems, loads program binary code into memory, and initiates execution of the application program, which then interacts with the user and with hardware devices. When a process is terminated, any reusable resources are released and returned to the OS.

Starting a new process is a heavy job for the OS, which includes allocating memory, creating data structures, and coping code.

A *thread* is a path of execution within a process and the basic unit to which the OS allocates processor time. A process can be *single-threaded* or *multithreaded*. Theoretically, a thread can do anything a process can do. The essential difference between a thread and a process is the work that each one is used to accomplish. Threads are used for small tasks, whereas processes are used for more heavyweight tasks – basically the execution of applications. Therefore, a thread is often called a *lightweight* process.

Threads within the same process share the same address space, whereas different processes do not. Threads also share global and static variables, file descriptors, signal bookkeeping, code area, and heap. This allows threads to read from and write to same data structures and variables and also facilitates communication between threads. Thus, threads use far less resources

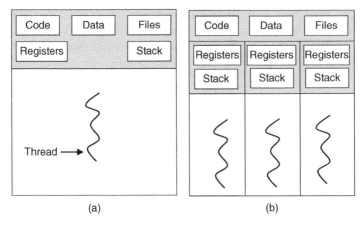

Figure 3.1 (a) Single-threaded process; (b) Multithreaded process.

compared to processes. However, each thread of the same process has its own thread status, program counter, registers, and stack, as illustrated in Figure 3.1.

A thread is the smallest unit of work managed independently by the scheduler of the OS. In RTOSs, the term *tasks* is often used for threads or single-threaded processes. For example, VxWorks and MicroC/OS-III are RTOSs that use the term tasks. In this book, processes (threads) and tasks are used interchangeably.

Processes may create other processes through appropriate system calls, such as *fork* or *spawn*. The process that does the creating is called the *parent* of the other process, which is called its *child*. Each process is assigned a unique integer identifier, call *process identifier*, or PID for short, when it is created. A process can be created with or without arguments. Generally, a parent process is in control of a child process. The parent can temporarily stop the child, cause it to be terminated, send it messages, look inside its memory, and so on. A child process may receive some amount of shared resources from its parent.

Processes may request their own termination by making the exit() system call. Processes may also be terminated by the system for a variety of reasons, such as the inability of the system to deliver necessary system resources, or in response to a *kill* command or other unhandled process interrupt. When a process terminates, all of its system resources are freed up, and open files are flushed and closed.

A process can be *suspended* for a variety of reasons. It can be *swapping* – the OS needs to release sufficient main memory to bring in a process that is ready to execute, or timing – a process that is executed periodically may be suspended while waiting for the next time interval. A parent process may also wish to suspend the execution of a child to examine or modify it or to coordinate the activity of various children.

Sometimes, processes need to communicate with each other while they are running. This is called *interprocess communication* (IPC). The OS provides mechanisms to support IPC. Files, sockets, message queues, pipes, named pipes, semaphores, shared memory, and message passing are typical IPC mechanisms.

3.1.2 Memory Management

Main memory is the most critical resource in a computer system in terms of speed at which programs run. The kernel of an OS is responsible for all system memory that is currently in use by programs. Entities in memory are data and instructions.

Each memory location has a *physical address*. In most computer architectures, memory is byte-addressable, meaning that data can be accessed 8 bits at a time, irrespective of the width of the data and address buses. Memory addresses are fixed-length sequences of digits. In general, only system software such as Basic Input/Output System (BIOS) and OS can address physical memory.

Most application programs do not have knowledge of physical addresses. Instead, they use logic addresses. A *logical address* is the address at which a memory location appears to reside from the perspective of an executing application program. A logical address may be different from the physical address due to the operation of an address translator or mapping function.

In a computer supporting *virtual memory*, the term physical address is used mostly to differentiate from a *virtual address*. In particular, in computers utilizing a *memory management unit* (MMU) to translate memory addresses, the virtual and physical addresses refer to an address before and after translation performed by the MMU, respectively. There are several reasons to use virtual memory. Among them is memory protection. If two or more processes are running at the same time and use direct addresses, a memory error in one process (e.g., reading a bad pointer) could destroy the memory being used by the other process, taking down multiple programs due to a single crash. The virtual memory technique, on the other hand, can ensure that each process is running in its own dedicated address space.

Before a program is loaded into memory by the *loader*, part of the OS, it must be converted into a *load module* and stored on disk. To create a load module, the source code is compiled by the compiler. The compiler produces an *object module*. An object module contains a header that records the size of each of the sections that follow, a machine code section that contains the executable instructions compiled by the compiler, an initialized data section that contains all data used by the program that require initialization, and the symbol table section that contains all external symbols used in the program. Some external symbols are defined in this object module and will be referred to by other object modules, and some are used in this object module and

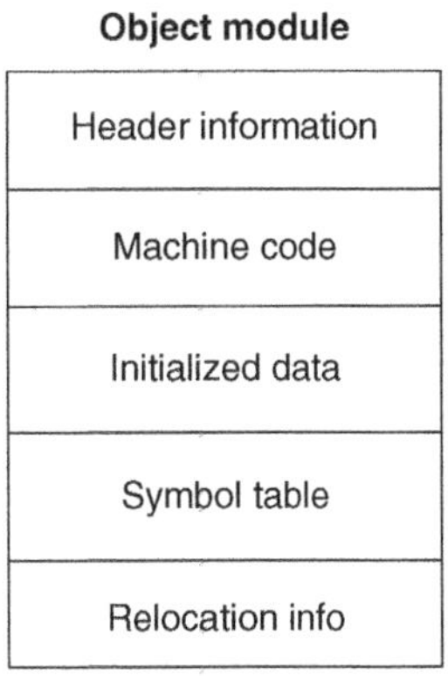

Figure 3.2 Structure of object module.

Object module

Header information
Machine code
Initialized data
Symbol table
Relocation info

defined in other object modules. The relocation information is used by the *linker* to combine several object modules into a load module. Figure 3.2 shows the structure of object modules.

When a process is started, the OS allocates memory to the process and then loads the load module from disk into the memory allocated to the process. During the loading, the executable code and initialized data are copied into the process' memory from the load module. In addition, memory is also allocated for uninitialized data and runtime stack that is used to keep information about each procedure call. The loader has a default initial size for the stack. When the stack is filled up at runtime, extra space will be allocated to it, as long as the predefined maximum size is not exceeded.

Many programming languages support memory allocation while a program is running. It is done by the call of *new* in C++ and Java or *malloc* in C, for example. This memory comes from a large pool of memory called the *heap* or *free store*. At any given time, some parts of the heap are in use, while some are free and thus available for future allocation. Figure 3.3 illustrates the memory areas of a running process, in which the heap area is the memory allocated by the process at runtime.

To avoid loading big executable files into memory, modern OS provides two services: *dynamic loading* and *dynamic linking*. In dynamic loading, a routine (library or other binary module) of a program is not loaded until it is called by the program. All routines are kept on disk in a relocatable load format. The main program is loaded into memory and is executed. Other routine methods or modules are loaded on request. Dynamic loading makes better memory space utilization, and unused routines are never loaded. Dynamic loading is useful when a large amount of code is needed to handle infrequently occurring cases.

In dynamic linking, libraries are linked at execution time. This is compared to *static linking*, in which libraries are linked at compile time, and thus, the resultant executable code is big. Dynamic linking refers to resolving

Process memory

Executable code
Initialized data
Uninitialized data
Heap
Stack

Figure 3.3 Memory areas of process.

symbols – associating their names with addresses or offsets – after compile time. It is particularly useful for libraries.

The simplest memory management technique is *single contiguous allocation*. That is, except the area reserved for the OS, all the memory is made available to a single application. This is the technique used in the MS-DOS system.

Another technique is *partitioned allocation*. It divides memory into multiple memory partitions; each partition can have only one process. The memory manager selects a free partition and assigns it to a process when the process starts and deallocates it when the process is finished. Some systems allow a partition to be *swapped* out to secondary storage to free additional memory. It is brought back into memory for continued execution later.

Paged allocation divides memory into fixed-size units called page frames and the program's *virtual address space* into pages of the same size. The hardware MMU maps pages to frames. The physical memory can be allocated on a page basis while the address space appears to be contiguous. Paging does not distinguish and protect programs and data separately.

Segmented memory allows programs and data to be broken up into logically independent address spaces and to aid sharing and protection. Segments are areas of memory that usually correspond to a logical grouping of information such as a code procedure or a data array. Segments require hardware support in the form of a segment table, which usually contains the physical address of the segment in memory, its size, and other data such as access protection bits and status (swapped in, swapped out, etc.).

As processes are loaded and removed from memory, the long, contiguous free memory space becomes fragmented into smaller and smaller contiguous pieces. Eventually, it may become impossible for the program to obtain large contiguous chunks of memory. The problem is called *fragmentation*. In general, smaller page size reduces fragmentation. The negative side is that it increases the page table size.

3.1.3 Interrupts Management

An *interrupt* is a signal from a device attached to a computer or from a running process within the computer, indicating an event that needs immediate attention. The processor responds by suspending its current activity, saving its state, and executing a function called an *interrupt handler* (also called *interrupt service routine*, ISR) to deal with the event.

Modern OSs are *interrupt-driven*. Virtually, all activities are initiated by the arrival of interrupts. Interrupt transfers control to the ISR, through the *interrupt vector*, which contains the addresses of all the service routines. Interrupt architecture must save the address of the interrupted instruction. Incoming interrupts are disabled while another interrupt is being processed. A system call is a software-generated interrupt caused either by an error or by a user request.

3.1.4 Multitasking

Real-world events may occur simultaneously. Multitasking refers to the capability of an OS that supports multiple independent programs running on the same computer. It is mainly achieved through time-sharing, which means that each program uses a share of the computer's time to execute. How to share processors' time among multiple tasks is addressed by *schedulers*, which follow scheduling algorithms to decide when to execute which task on which processor.

Each task has a *context*, which is the data indicating its execution state and stored in the *task control block* (TCB), a data structure that contains all the information that is pertinent to the execution of the task. When a scheduler switches a task out of the CPU, its context has to be stored; when the task is selected to run again, the task's context is restored so that the task can be executed from the last interrupted point. The process of storing and restoring the context of a task during a task switch is called a *context switch*, which is illustrated in Figure 3.4.

Context switches are the overhead of multitasking. They are usually computationally intensive. Context switch optimization is one of the tasks of OS design. This is particularly the case in RTOS design.

3.1.5 File System Management

Files are the fundamental abstraction of secondary storage devices. Each file is a named collection of data stored in a device. An important component of an OS is the *file system*, which provides capabilities of file management, auxiliary storage management, file access control, and integrity assurance.

File management is concerned with providing the mechanisms for files to be stored, referenced, shared, and secured. When a file is created, the file system

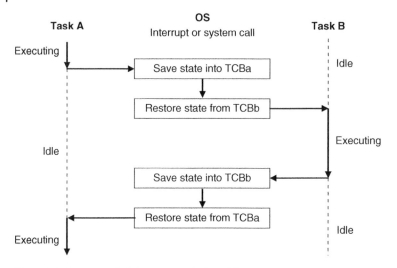

Figure 3.4 Context switch between tasks A and B.

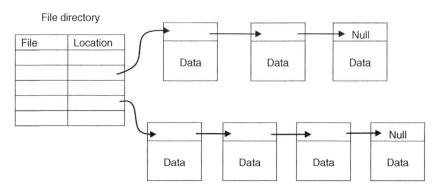

Figure 3.5 The block chaining scheme in file storage.

allocates an initial space for the data. Subsequent incremental allocations follow as the file grows. When a file is deleted or its size is shrunk, the space that is freed up is considered available for use by other files. This creates alternating used and unused areas of various sizes. When a file is created and there is not an area of contiguous space available for its initial allocation, the space must be assigned in *fragments*. Because files do tend to grow or shrink over time, and because users rarely know in advance how large their files will be, it makes sense to adopt noncontiguous storage allocation schemes. Figure 3.5 illustrates the block chaining scheme. The initial address of storage of a file is identified by its file name.

Typically, files on a computer are organized into directories, which constitute a hierarchical system of tree structure.

A file system typically stores necessary bookkeeping information for each file, including the size of data contained in the file, the time the file was last modified, its owner user ID and group ID, and its access permissions.

The file system also provides a spectrum of commands to read and write the contents of a file, to set the file read/write position, to set and use the protection mechanism, to change the ownership, to list files in a directory, and to remove a file.

File access control can be realized using a two-dimensional matrix that lists all users and all files in the system. The entry at index (i, j) of the matrix specifies if user i is allowed to access file j. In a system that has a large number of users and contains a large number of files, this matrix would be very large and very sparse.

A scheme that requires much less space is to control access to various user classes. *Role-based access control* (RBAC) is an access control method where access to data is performed by authorized users. RBAC assigns users to specific roles, and permissions are granted to each role based on the user's job requirements. Users can be assigned a number of roles in order to conduct day-to-day tasks. For example, a user may need to have a developer role as well as an analyst role. Each role would define the permissions that are needed to access different objects.

3.1.6 I/O Management

Modern computers interact with a wide range of I/O devices. Keyboards, mice, printers, disk drives, USB drives, monitors, networking adapters, and audio systems are among the most common ones. One purpose of an OS is to hide peculiarities of hardware I/O devices from the user.

In *memory-mapped I/O*, each I/O device occupies some locations in the I/O address space. Communication between the I/O device and the processor is enabled through physical memory locations in the I/O address space. By reading from or writing to those addresses, the processor gets information from or sends commands to I/O devices.

Most systems use *device controllers*. A device controller is primarily an interface unit. The OS communicates with the I/O device through the device controller. Nearly all device controllers have *direct memory access* (DMA) capability, meaning that they can directly access the memory in the system, without the intervention by the processor. This frees up the processor of the burden of data transfer from and to I/O devices.

Interrupts allow devices to notify the processor when they have data to transfer or when an operation is complete, allowing the processor to perform other duties when no I/O transfers need its immediate attention. The processor has the interrupt request line that it senses after executing every instruction. When a device controller raises an interrupt by asserting a signal on the interrupt request line, the processor catches it and saves the state and then transfers control to the interrupt handler. The interrupt handler determines the cause of the interrupt, performs necessary processing, and executes a *return from interrupt* instruction to return control to the processor.

I/O operations often have high latencies. Most of this latency is due to the slow speed of peripheral devices. For example, information cannot be read from or written to a hard disk until the spinning of the disk brings the target sectors directly under the read/write head. The latency can be alleviated by having one or more input and output *buffers* associated with each device.

3.2 Characteristics of RTOS Kernels

Although a general-purpose OS provides a rich set of services that are also needed by real-time systems, it takes too much space and contains too many functions that may not be necessary for a specific real-time application. Moreover, it is not configurable, and its inherent timing uncertainty offers no guarantee to system response time. Therefore, a general-purpose OS is not suitable for real-time embedded systems.

There are three key requirements of RTOS design. Firstly, the timing behavior of the OS must be predictable. For all services provided by the OS, the upper bound on the execution time must be known. Some of these services include OS calls and interrupt handling. Secondly, the OS must manage the timing and scheduling, and the scheduler must be aware of task deadlines. Thirdly, the OS must be fast. For example, the overhead of context switch should be short. A fast RTOS helps take care of soft real-time constraints of a system as well as guarantees hard deadlines.

As illustrated in Figure 3.6, an RTOS generally contains a *real-time kernel* and other higher-level services such as file management, protocol stacks, a Graphical User Interface (GUI), and other components. Most additional services revolve around I/O devices. A real-time kernel is software that manages the time and resources of a microprocessor or microcontroller and provides indispensable services such as task scheduling and interrupt handling to applications. Figure 3.7 shows a general structure of a microkernel. In embedded systems, a small amount of code called *board support package* (BSP) is implemented for a given board that conforms to a given OS. It is commonly built with a bootloader that contains the minimal device support to load the OS and device drivers for all the devices on the board.

In the rest of this section, we introduce some most important real-time services that are specified in POSIX 1.b for RTOS kernels.

3.2.1 Clocks and Timers

Most embedded systems must keep track of the passage of time. The length of time is represented by the number of system ticks in most RTOS kernels.

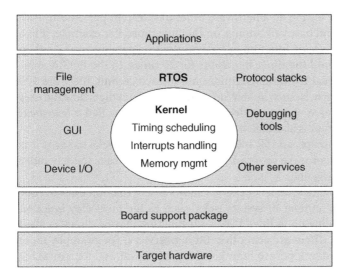

Figure 3.6 A high-level view of RTOS.

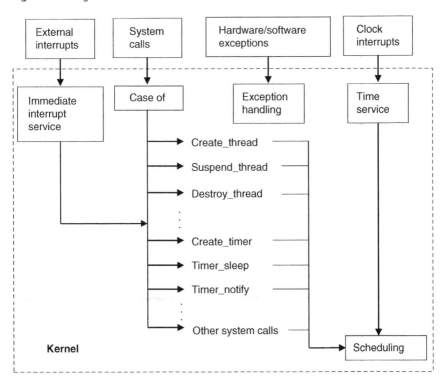

Figure 3.7 Structure of a microkernel.

The RTOS works by setting up a hardware timer to interrupt periodically, say, every millisecond, and bases all timings on the interrupts. For example, if in a task you call the *taskDelay* function in VxWorks with a parameter of 20, then the task will block until the timer interrupts for 20 times. In the POSIX standard, each tick is equal to 10 milliseconds, and in each second, there are 100 ticks. Some RTOS kernels, such as VxWorks, define routines that allow the user to set and get the value of system tick. The timer is often called a *heartbeat timer*, and the interrupt is also called *clock interrupt*.

At each clock interrupt, an ISR increments tick count, checks to see if it is now time to unblock or wake up a task. If this is the case, it calls the scheduler to do the scheduling again.

Based on the system tick, an RTOS kernel allows you to call functions of your choice after a given number of system ticks, such as the *taskDelay* function in VxWorks. Depending upon the RTOS, your function may be directly called from the timer ISR. There are also other timing services. For example, most RTOS kernels allow developers to limit how long a task will wait for a message from a queue or a mailbox and how long a task will wait for a semaphore, and so on.

Timers improve the determinism of real-time applications. Timers allow applications to set up events at predefined intervals or time. POSIX specified several timer-related routines, including the following:

- `timer_create()` – allocate a timer using the specified clock for a timing base.
- `timer_delete()` – remove a previously created timer.
- `timer_gettime()` – get the remaining time before expiration and the reload value.
- `timer_getoverrun()` – return the timer expiration overrun.
- `timer_settime()` – set the time until the next expiration and arm timer.
- `nanosleep()` – suspend the current task until the time interval elapses.

3.2.2 Priority Scheduling

Because real-time tasks have deadlines, being soft or hard, all tasks are not equal in terms of the urgency of getting executed. Tasks with shorter deadlines should be scheduled for execution sooner than those with longer deadlines. Therefore, tasks are typically prioritized in an RTOS. Moreover, if a higher priority task is released while the processor is serving a lower priority task, the RTOS should temporarily suspend the lower priority task and immediately schedule the higher priority on the processor for execution, to ensure that the higher priority task is executed before its deadline. This process is called *preemption*. Task scheduling for real-time applications is typically priority-based and preemptive. Examples are earliest deadline first (EDF) scheduling and rate monotonic (RM) scheduling. Scheduling algorithms that do not take task

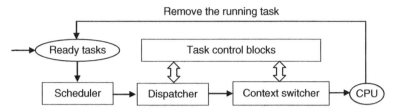

Figure 3.8 Priority scheduling.

priorities into account, such as first-in-first-service and round-robin, are not suitable for real-time systems.

In priority-driven preemptive scheduling, the preemptive scheduler has a clock interrupt task that provides the scheduler with options to switch after the task has had a given period to execute – the time slice. This scheduling system has the advantage of making sure that no task hogs the processor for any time longer than the time slice.

As shown in Figure 3.8, an important component involved in scheduling is the *dispatcher*, which is the module that gives control of the CPU to the task selected by the scheduler. It receives control in kernel mode as the result of an interrupt or system call. It is responsible for performing context switches. The dispatcher should be as fast as possible, since it is invoked during every task switch. During the context switches, the processor is virtually idle for a fraction of time; thus, unnecessary context switches should be avoided.

The key to the performance of priority-driven scheduling is in choosing priorities for tasks. Priority-driven scheduling may cause low-priority tasks to starve and miss their deadlines. In the next two chapters, several well-known scheduling algorithms and resource access control protocols will be discussed.

3.2.3 Intertask Communication and Resource Sharing

In an RTOS, a task cannot call another task. Instead, tasks exchange information through message passing or memory sharing and coordinate their execution and access to shared data using real-time signals, mutex, or semaphore objects.

3.2.3.1 Real-Time Signals

Signals are similar to software interrupts. In an RTOS, a signal is automatically sent to the parent when a child process terminates. Signals are also used for many other synchronous and asynchronous notifications, such as waking a process when a wait call is performed and informing a process that it has issued a memory violation.

POSIX extended the signal generation and delivery to improve the real-time capabilities. Signals take an important role in real-time systems as the way to inform the processes of the occurrence of asynchronous events such as high-resolution timer expiration, fast interprocess message arrival, asynchronous I/O completion, and explicit signal delivery.

3.2.3.2 Semaphores

Semaphores are counters used for controlling access to resources shared among processes or threads. The value of a semaphore is the number of units of the resource that are currently available. There are two basic operations on semaphores. One is to atomically increment the counter. The other is to wait until the counter is nonnull and atomically decrement it. A semaphore tracks only how many resources are free; it does not keep track of which of the resources are free.

A binary semaphore acts similarly to a *mutex*, which is used in the case when a resource can only be used by at most one task at any time.

3.2.3.3 Message Passing

In addition to signals and semaphores, tasks can share data by sending messages in an organized *message passing* scheme. Message passing is much more useful for information transfer. It can also be used just for synchronization. Message passing often coexists with shared memory communication. Message contents can be anything that is mutually comprehensible between the two parties in communication. Two basic operations are *send* and *receive*.

Message passing can be *direct* or *indirect*. In direct message passing, each process wanting to communicate must explicitly name the recipient or sender of the communication. In indirect message passing, messages are sent to and received from *mailboxes* or *ports*. Two processes can communicate in this way only if they have a shared mailbox.

Message passing can also be *synchronous* or *asynchronous*. In synchronous message passing, the sender process is blocked until the message primitive has been performed. In asynchronous message passing, the sender process immediately gets control back.

3.2.3.4 Shared Memory

Shared memory is a method that an RTOS uses to map common physical space into independent process-specific virtual space. Shared memory is commonly used to share information (resources) between different processes or threads. Shared memory must be accessed exclusively. Therefore, it is necessary to use mutex or semaphores to protect the memory area. The code segment in a task that accesses the shared data is called a *critical section*. Figure 3.9 shows that two tasks share a memory region.

A side effect in using shared memory is that it may cause *priority inversion*, a situation that a low-priority task is running while a high-priority task is waiting. More details of priority inversion will be discussed in Chapter 5.

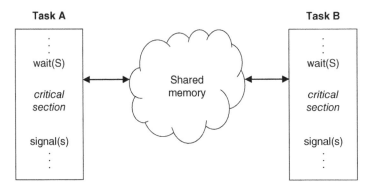

Figure 3.9 Shared memory and critical section.

3.2.4 Asynchronous I/O

There are two types of I/O synchronization: synchronous I/O and asynchronous I/O. In synchronous I/O, when a user task requests the kernel for I/O operation and the requested is granted, the system will wait until the operation is completed before it can process other tasks. Synchronous I/O is desirable when the I/O operation is fast. It is also easy to implement.

An RTOS supports the overlap of application processing and application-initiated I/O operations. This is the RTOS service of asynchronous I/O. In asynchronous I/O, after a task requests I/O operation, while this task is waiting for I/O to complete, other tasks that do not depend on the I/O results will be scheduled for execution. Meanwhile, tasks that depend on the I/O having completed are blocked. Asynchronous I/O is used to improve throughput, latency, and/or responsiveness.

Figure 3.10 illustrates the idea of synchronous I/O and asynchronous I/O.

3.2.5 Memory Locking

Memory locking is a real-time capability specified by POSIX that is intended for a process to avoid the latency of fetching a page of memory. It is achieved by locking the memory so that the page is *memory-resident*, that is, it remains in the main memory. This allows fine-grained control of which part of the application must stay in physical memory to reduce the overhead associated with transferring data between memory and disk. For example, memory locking can be used to keep in memory a thread that monitors a critical process that requires immediate attention.

When the process exits, the locked memory is automatically unlocked. Locked memory can also be unlocked explicitly. POSIX, for example, defined mlock() and munlock() functions to lock and unlock memory. The munlock function unlocks the specified address range regardless of the number of times the mlock function was called. In other words, you can lock

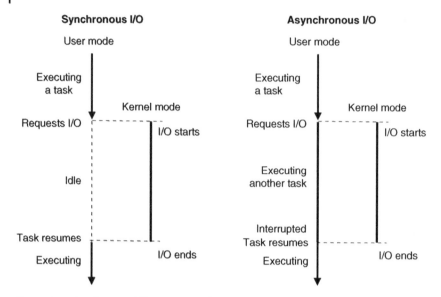

Figure 3.10 Synchronous I/O versus asynchronous I/O.

address ranges over multiple calls to the mlock function, but the locks can be removed with a single call to the munlock function. In other words, memory locks don't stack.

More than one process can lock the same or overlapping region, and in that case, the memory region remains locked until all the processes have unlocked it.

3.3 RTOS Examples

3.3.1 LynxOS

The LynxOS RTOS is a Unix-like RTOS from Lynx Software Technologies. LynxOS is a deterministic, hard RTOS that provides POSIX-conformant APIs in a small-footprint embedded kernel. It features predictable worst-case response time, preemptive scheduling, real-time priorities, ROMable kernel, and memory locking. LynxOS provides symmetric multiprocessing support to fully take advantage of multicore/multithreaded processors. LynxOS 7.0, the latest version, includes new tool chains, debuggers, and cross-development host support.

The LynxOS RTOS is designed from the ground up for conformance to open-system interfaces. It leverages existing Linux, UNIX, and POSIX programming talent for embedded real-time projects. Real-time system development time is saved, and programmers are able to be more productive using familiar methodologies as opposed to learning proprietary methods.

LynxOS is mostly used in real-time embedded systems, in applications for avionics, aerospace, the military, industrial process control, and telecommunications. LynxOS has already been used in millions of devices.

3.3.2 OSE

OSE is an acronym for the operating system embedded. It is a real-time embedded OS created by the Swedish information technology company ENEA AB. OSE uses signals in the form of messages passed to and from processes in the system. Messages are stored in a queue attached to each process. A link handler mechanism allows signals to be passed between processes on separate machines, over a variety of transports. The OSE signaling mechanism formed the basis of an open-source interprocess kernel design.

The Enea RTOS family shares a high-level programming model and an intuitive API to simplify programming. It consists of two products, each optimized for a specific class of applications:

- Enea OSE is a robust and high-performance RTOS optimized for multicore, distributed, and fault-tolerant systems.
- Enea OSEck is a compact and multicore DSP-optimized version of ENEA's full-featured OSE RTOS.

OSE supports many mainly 32-bit processors, such as those in ARM, PowerPC, and MIPS families.

3.3.3 QNX

The QNX Neutrino RTOS is a full-featured and robust OS developed by QNX Software Systems Limited, a subsidiary of BlackBerry. QNX products are designed for embedded systems running on various platforms, including ARM and x86, and a host of boards implemented in virtually every type of embedded environment.

As a microkernel-based OS, QNX is based on the idea of running most of the OS kernel in the form of a number of small tasks, known as servers. This differs from the more traditional monolithic kernel, in which the OS kernel is a single, very large program composed of a huge number of components with special abilities. In the case of QNX, the use of a microkernel allows developers to turn off any functionality they do not require without having to change the OS itself; instead, those servers will simply not run.

The BlackBerry PlayBook tablet computer designed by BlackBerry uses a version of QNX as the primary OS. The BlackBerry line of devices running the BlackBerry 10 OS are also based on QNX.

3.3.4 VxWorks

VxWorks is an RTOS developed as proprietary software by Wind River, a subsidiary company of Intel providing embedded system software, which comprises runtime software, industry-specific software solutions, simulation technology, development tools, and middleware. As other RTOS products, VxWorks is designed for use in embedded systems requiring real-time and deterministic performance.

VxWorks supports Intel, MIPS, PowerPC, SH-4, and ARM architectures. The VxWorks Core Platform consists of a set of runtime components and development tools. VxWorks core development tools are compilers such as Diab, GNU, and Intel C++ Compiler (ICC) and its build and configuration tools. The system also includes productivity tools such as its Workbench development suite and Intel tools and development support tools for asset tracking and host support. Cross-compiling is used with VxWorks. Development is achieved on a "host" system where an integrated development environment (IDE), including the editor, compiler toolchain, debugger, and emulator, can be used. Software is then compiled to run on the "target" system. This allows the developer to work with powerful development tools while targeting more limited hardware.

VxWorks is used by products over a wide range of market areas: aerospace and defenses, automotive, industrial such as robots, consumer electronics, medical area, and networking. Several notable products also use VxWorks as the onboard OS. Examples in the area of spacecraft are the Mars Reconnaissance Orbiter, the Phoenix Mars Lander, the Deep Impact Space Probe, and the Mars Pathfinder.

3.3.5 Windows Embedded Compact

Formerly known as Windows CE, Windows Embedded Compact is an OS subfamily developed by Microsoft as part of its Windows Embedded family of products. It is a small-footprint RTOS and optimized for devices that have minimal memory, such as industrial controllers, communication hubs, and consumer electronics devices such as digital cameras, GPS systems, and also automotive infotainment systems. It supports x86, SH (automotive only), and ARM.

Exercises

1 What is the kernel of an operating system? In which mode does it run?

2 What are the two approaches that a user process interacts with the operating system? Discuss the merits of each approach.

3 What is the difference between a program and a process? What is the difference between a process and a thread?

4 Should terminating a process also terminate all its children? Give an example when this is a good idea and another example when this is a bad idea.

5 Should the open files of a process be automatically closed when the process exits?

6 What is the difference between an object module and a load module?

7 Discuss the merits of each memory allocation techniques introduced in this chapter.

8 What do we mean when we say an OS is interrupt-driven? What does the processor do when an interrupt occurs?

9 What is a context switch and when does it occur?

10 Does a file have to be stored in a contiguous storage region in a disk?

11 What are the benefits of using memory-mapped I/O?

12 Why can a general-purpose OS not meet the requirements of real-time systems?

13 How does memory fragmentation form? Name one approach to control it.

14 What are the basic functions of an RTOS kernel?

15 How does an RTOS keep track of the passage of time?

16 Why is it necessary to use priority-based scheduling in a real-time application?

17 What are the general approaches that different tasks communicate with each other and perform synchronization of their actions in access to shared resources?

18 Compare synchronous I/O and asynchronous I/O and list their advantages and disadvantages.

19 How does the memory locking technique improve the performance of a real-time system?

Suggestions for Reading

Many textbooks are available that provide thorough and in-depth discussion on general-purpose OSs. Examples are [1–3]. Cooling [4] provides a great overview of the fundamentals of RTOS for embedded programming without being specific to any one vendor. POSIX-specified real-time facilities are described in Ref. [5]. More information regarding the commercial RTOS products can be found on their official websites.

References

1 Doeppner, T. (2011) *Operating Systems in Depth*, Wiley.
2 McHoes, A.M. and Flynn, I.M. (2011) *Understand Operating Systems*, 6th edn, Course Technology Cengage Learning.
3 Stallings, W. (2014) Operating Systems: Internals and Design Principles, 8th edn, Pearson.
4 Cooling, J. (2013) *Real-Time Operating Systems*, Kindle edn, Lindentree Associates.
5 Gallmeister, B. (1995) POSIX 4.0: Programming for the Real World, O'Reilly & Associates, Inc..

URLs

LynxOS, http://www.lynxos.com/
VxWorks, http://www.windriver.com/products/vxworks/
Windows CE, https://www.microsoft.com/windowsembedded/en-us/windows-embedded-compact-7.aspx
QNX, http://www.qnx.com/content/qnx/en.html
OSE, http://www.enea.com/ose

4

Task Scheduling

Task management and scheduling are some of the core functions of any RTOS kernel. Part of the kernel is a scheduler that is responsible for allocating and scheduling tasks on processors to ensure that deadlines are met. This chapter presents some well-known and widely used techniques for task assignment and scheduling.

4.1 Tasks

A *task* is a unit of work scheduled for execution on the CPU. It is a building block of real-time application software supported by an RTOS. In fact, a real-time application that uses an RTOS can be structured as a set of independent tasks. There are three types of tasks:

Periodic tasks. Periodic tasks are repeated once a period, for example, 200 milliseconds. They are time-driven. Periodic tasks typically arise from sensory data acquisition, control law computation, action planning, and system monitoring. Such activities need to be cyclically executed at specific rates, which can be derived from the application requirements. Periodic tasks have hard deadlines, because each instance of a periodic task has to complete execution before the next instance is released. Otherwise, task instances will pile up.

Aperiodic tasks. Aperiodic tasks are one-shot tasks. They are event-driven. For example, a driver may change the vehicle's cruise speed while the cruise control system is in operation. To maintain the speed set by the driver, the system periodically takes its speed signal from a rotating driveshaft, speedometer cable, wheel speed sensor, or internal speed pulses produced electronically by the vehicle and then pulls the throttle cable with a solenoid as needed. When the user manually changes the speed, the system has to respond to the change and meanwhile keeps its regular operation. Aperiodic tasks either have no deadlines or have soft deadlines.

Real-Time Embedded Systems, First Edition. Jiacun Wang.
© 2017 John Wiley & Sons, Inc. Published 2017 by John Wiley & Sons, Inc.

Sporadic tasks. Sporadic tasks are also event-driven. The arrival times of sporadic task instances are not known *a priori*, but there is requirement on the minimum interarrival time. Unlike aperiodic tasks that do not have hard deadlines, sporadic tasks have hard deadlines. For example, when the driver of a vehicle sees a dangerous situation in front of him and pushes the break to stop the vehicle, the speed control system has to respond to the event (a hard step on the break) within a small time window.

4.1.1 Task Specification

In a real-time system, a task can be specified by the following temporal parameters:

Release time. The release time of task is the time when a task becomes available for execution. The task can be scheduled for execution at any time at or after the release time. It may not be executed immediately, because, for example, a higher or equal-priority task is using the processor. The release time of a task T_i is denoted by r_i.

Deadline. The deadline of a task is the instant of time by which its execution must be completed. The deadline of T_i is denoted by d_i.

Relative deadline. The relative deadline of a task is the deadline measured in reference to its release time. For example, if a task is released at time t and its deadline is $t + 200$ milliseconds, then its relative deadline is 200 milliseconds. The relative deadline of T_i is denoted by D_i.

Execution time. The execution time of a task is the amount of time that is required to complete the execution of the task when it executes alone and has all required resources in place. A task's execution time mainly depends on the complexity of the task and the speed of the processor. The execution time of T_i is denoted by e_i.

Response time. The response time of a task is the length of time elapsed from the task is released to the execution is completed. For a task with a hard deadline, the maximum allowed response time is the task's relative deadline.

Figure 4.1 illustrates these five temporal parameters. Real-time constraints imposed on a task are typically specified in terms of its release time and deadline.

In addition to the aforementioned parameters, a periodic task has the following two parameters:

Period. The period of a periodic task is the length of the interval between the release times of two consecutive instances of the task. We assume that all intervals have the same length throughout the book. The period of T_i is denoted by p_i.

Phase. The phase of a periodic task is the release time of its first instance. The phase of T_i is denoted by ϕ_i.

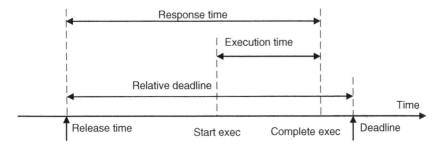

Figure 4.1 Task specification.

Utilization. The utilization of a periodic task is the ratio of its execution time over its period, denoted by u_i. $u_i = e_i/p_i$.

For a periodic task, the task execution time, release time, deadline, relative deadline, and response time are all referring to its instances. We assume that all instances of a periodic task have the same execution time throughout the book. We specify a periodic task as follows:

$$T_i = (\phi_i, p_i, e_i.D_i)$$

For example, a task with parameters (2, 10, 3, 9) would mean that the first instance of the task is released at time 2, the following instances will arrive at 12, 22, 32, and so on. The execution time of each instance is 3 units of time. When an instance is released, it has to be executed within 9 units of time.

If the task's phase is 0, then we specify it with

$$T_i = (p_i, e_i.D_i).$$

If the task's relative deadline is the same as its period, then we specify it with two parameters only:

$$T_i = (p_i, e_i).$$

Given a set of periodic tasks $T_i, i = 1, 2, ..., n$, we can calculate their *hyperperiod*, denoted by H. H is the least common multiple (LCM) of p_i for $i = 1, 2, ..., n$. One way to calculate H is prime factorization. The theory behind this approach is that every positive integer greater than 1 can be written in only one way as a product of prime numbers. The prime numbers can be considered as the atomic elements, which, when combined together, make up a composite number.

Example 4.1 *Calculation of Hyperperiod of Periodic Tasks*
Consider a system of three periodic tasks:

$$T_1 = (5, 1), T_2 = (12, 2), T_3 = (40, 1).$$

Number 5 is a prime and is not decomposable. Number 12 is a composite number and can be expressed as

$$12 = 2 \cdot 6 = 2 \cdot 2 \cdot 3 = 2^2 \cdot 3.$$

Number 40 is also a composite number and can be decomposed as

$$40 = 2 \cdot 20 = 2 \cdot 2 \cdot 10 = 2 \cdot 2 \cdot 2 \cdot 5 = 2^3 \cdot 5.$$

The LCM of 5, 12, and 20 is the product of the highest power of each prime number combined. That is,

$$H = \mathrm{LCM}(5, 12, 40)$$
$$= 2^3 \cdot 3 \cdot 5$$
$$= 120$$

When we compute the schedule for a set of periodic tasks, we only need to compute its schedule for its first hyperperiod, and then, we reuse the same schedule for each hyperperiod afterward.

In addition to timing parameters, tasks possess functional parameters that are important in task scheduling.

Criticality. Tasks in a system are not equally important. The relative priorities of tasks are a function of the nature of the tasks themselves and the current state of the controlled process. The priority of a task indicates the criticalness of the task with respect to other tasks.

Preemptivity. Execution of tasks can be interleaved. The scheduler may suspend the execution of a task and give the processor to a more urgent task. The suspended task may resume its execution when the more urgent task completed. Such an interrupt of task execution is called a *preemption*. A task is *preemptable* if it can resume its execution from the point of interruption. In other words, it does not need to start over. An example is a computational task on the CPU. On the other hand, a task is *nonpreemptable* if it must be executed from start to completion without interruption. If they are interrupted in the middle of execution, they have to be executed again from the very beginning.

A task can be *partially preemptive.* For example, if part of a task accesses exclusively shared resources, then the critical section is nonpreemptable, but the rest of the task may be preemptable.

4.1.2 Task States

In real-time systems, a task can exist in one of the following three states:

Running. When a task is actually executing, it is said to be in the *running* state. It is currently utilizing the processor. In a single-processor system, only one task can be in the *running* state at any given time.

Ready. Tasks in the *ready* state are those that are able to execute but are not currently executing because a different task of equal or higher priority is in the *running* state. Ready tasks have all resources needed to run except the processor. In this state, a task is able to run whenever it becomes the task with the highest priority among all tasks in this state and the processor is released. Any number of tasks can be in this state.

Blocked. A task is said to be in the *blocked* state if it is currently waiting for either a temporal or an external event. For example, if a task calls *taskDelay()*, it will block itself until the delay period has expired – timer expiration is a temporal event. A task that is responsible for processing user inputs has nothing to do until the user types in something, a case of an external event. Tasks can also block while they are waiting for RTOS kernel object events. Tasks in the *blocked* state are not available for scheduling. Any number of tasks can be in this state as well.

When a new task is created, it is placed in the ready state queue. The task can be scheduled for execution immediately or later, depending on it priority and the priority of other tasks in the *ready* state. All tasks in this state complete for the processor to run. When a task with the highest priority is dispatched for execution on the processor, it shifts to the *running* state.

If it is preemptable and the scheduler is priority–preemption based, a task in the *running* state may be preempted by a higher priority task. When it is preempted, the RTOS kernel places it in the *ready* queue. A task in the *running* state may also enter the *blocked* state for several reasons. Let us explain the blocking case that causes priority inversion here. Assume that two tasks, namely *A* and *B*, share a piece of common memory that is designed to be used exclusively. *A* has a higher priority over *B*. While *B* is running and accessing the shared memory, *A* is released and preempts *B*. When *A* starts to execute its code that accesses the shared memory, it is blocked because the shared memory is not available. Now *B* in the *ready* state is dispatched to run. This is the situation that a lower priority task is running while a higher priority task is waiting.

A task is blocked because all conditions other than the processor for its execution are not met. Examples of the conditions are time delay and required resources. When all the conditions are met, the task will be placed in the *ready* state. In the priority inversion example, when the task *B* exits the shared memory access, it will signal the RTOS kernel. The task *A* will preempt *B* immediately and resume its execution.

Figure 4.2 describes the state transitions of a task. Notice that some RTOS kernels defined more states for tasks. For example, the T-Kernel RTOS defined five task states: *running, ready, waiting, suspended,* and *waiting-suspended.* In addition to the *running* and *ready* states, a task in VxWorks can be in the state of *pending, suspended, delayed,* or a combination of these states.

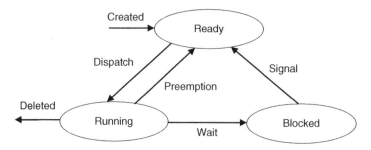

Figure 4.2 States of a task.

4.1.3 Precedence Constraints

In addition to timing constraints on tasks in a real-time application, there might be *precedence constraints* among tasks. A precedence constraint specifies the execution order of two or more tasks. It reflects data and/or control dependencies among tasks. For example, in a real-time control system, the control law computation task has to wait for the results from the sensor data polling task. Therefore, the sensor data pulling task should run before the control law computation task in each control cycle.

If a task A is an *immediate* predecessor of a task B, we use $A < B$ to represent their precedence relation. To be more intuitive, we can use a *task graph* to show the precedence constraints among a set of tasks. Nodes of the graph are tasks, and a directed edge from the node A to the node B indicates that the task A is an immediate predecessor of the task B.

Example 4.2 *Precedence Graph*
There are seven tasks shown in Figure 4.3. Dependency constraints among them are as follows:

$$T_1 < T_3, T_2 < T_3, T_3 < T_6, T_4 < T_5, T_5 < T_6, \text{ and } T_6 < T_7.$$

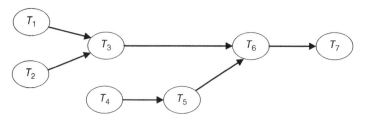

Figure 4.3 A task graph.

4.1.4   Task Assignment and Scheduling

Given all task timing parameters and functional parameters, the tasks are allocated or assigned to the various processors and scheduled on them. Assigning and scheduling problems with multiple processors are NP-complete, and thus, heuristic approaches are often employed. Developing a multiprocessor schedule normally proceeds with two steps: we first assign tasks to processors, and then, we do a uniprocessor scheduling for tasks assigned to each processor. If one or more of the schedules are infeasible, then we either redo the task allocation or use a different algorithm to do the scheduling. It is possible that there is no feasible assignment/schedule for a particular problem. Notice that all scheduling algorithms presented in this chapter are uniprocessor-based. Task assignment techniques are introduced later this chapter.

Each task executes within its own context. Only one task within the application can be executing at any point in time on a processor. A *schedule* is an assignment of tasks to available processors. The *scheduler* of an RTOS is a module that implements the assignment (scheduling) algorithms.

A schedule is said to be *valid* if all precedence and resource usage constraints are met and no task is *underscheduled* (the task is assigned insufficient time for execution) or overscheduled (the task is assigned more time than needed for execution).

A schedule is said to be *feasible* if every task scheduled completes before its deadline. A system (a set of tasks) is said to be *schedulable* if a feasible schedule exists for the system. The bulk of real-time scheduling work deals with finding feasible schedules.

An algorithm is said to be *optimal* if it always produces a feasible schedule as long as a given set of tasks has feasible schedules.

We use *schedulable utilization* (SU) to measure the performance of a scheduling algorithm. The SU is the maximum total utilization of tasks that can be feasibly scheduled by the algorithm. The higher the SU, the better the scheduling algorithm.

A schedule may be computed before the system is put in operation or obtained dynamically at runtime.

4.2   Clock-Driven Scheduling

In the clock-driven scheduling, scheduling decisions are made at specific time instants, which are typically chosen *a priori* before the system begins execution. It usually works for *deterministic systems* where tasks have hard deadlines and all task parameters are not changing during system operation. Clock-driven schedules can be computed off-line and stored for use at runtime, which significantly saves runtime scheduling overhead.

For a set of periodic tasks with given parameters, it is quite straightforward to draw a feasible schedule for the tasks based on the clock-driven approach, as long as such a feasible schedule exists. In most cases, one can come up with different feasible schedules using this approach. It is up to the user to select the best one according to some criteria.

Example 4.3   *Clock-Driven Scheduling*

Figure 4.4 shows two feasible schedules for the system of three periodic tasks:

$$T_1 = (4, 1), T_2 = (6, 1), T_3 = (12, 2).$$

The schedules are in the interval from 0 to 12 units of time, the first hyperperiod of the system. During this time period, there are three instances from T_1, two instances from T_2, and one instance from T_3. Let us analyze the first schedule and see why it is feasible.

Consider T_1 first. At time 0, T_1 has its first instance released with a deadline at time 4. It is scheduled for execution at time 0 and completed at time 1. So, it meets its deadline.

At time 4, the second instance of T_1 is released with a deadline at time 8. It is scheduled for execution at time 4 and completed at time 5. It meets its deadline.

At time 8, the third instance of T_1 is released with a deadline at time 12. It is scheduled for execution at time 8 and completed at time 9. It meets its deadline.

Now let us consider T_2. T_2 has its first instance released at time 0. Its deadline is at time 6. It is scheduled for execution at time 1 and completed at time 2. So, it meets its deadline.

At time 6, the second instance of T_2 is released with a deadline at time 12. It is scheduled for execution at time 6 and completed at time 7. It meets its deadline.

The only instance in T_3 is released at time 0 with a deadline at time 12. It is scheduled for execution at time 2 and completed at time 4. So, it meets its deadline.

Since all task instances released in the first hyperperiod of the system meet their deadlines, the schedule is feasible.

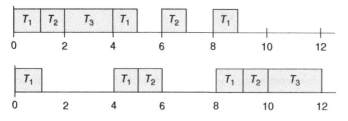

Figure 4.4 Two feasible schedules for $T_1 = (4, 1)$, $T_2 = (6, 1)$, and $T_3 = (12, 2)$.

A similar analysis will conclude that the second schedule is also feasible. In fact, one can move around the task execution slots a bit and get other feasible schedules. What is important is that you have to consider all instances in the time period and make sure that they all meet their deadlines.

You might wonder that why we only draw the schedule up to time 12. Well, the reason is that the hyperperiod of these three tasks is 12. In the clock-driven scheduling, the hyperperiod of a set of tasks is also called their *major cycle*. It can be proved that regardless of what each task's phase is, their major cycle is always their LCM.

We can use a schedule table to list all scheduling decisions in a schedule. For example, Table 4.1 is such a table for the first schedule shown in Figure 4.4. There are six task instances in the table. Let $N = 6$. Then based on the table, a scheduler can be designed according to the pseudocode shown in Figure 4.5.

Table 4.1 The schedule table for the first schedule shown in Figure 4.4.

Entry k	Time t_k	Task $T(t_k)$
0	0	T_1
1	1	T_2
2	2	T_3
3	4	T_1
4	6	T_2
5	8	T_1

```
Input: Stored scheduling table (t_k, T(t_k)) for k = 0, 1, ..., N-1.
Task SCHEDULER:
        i := 0;  //decision point
        k := 0;
        set timer to expire at t_k;
        loop FOREVER
                accept timer interrupt;
                currentTask := T(t_k);
                i++;
                k := i mod N;
                set timer to expire at floor(i/N)H+ t_k;
                execute currentTask;
                sleep;
        end loop
end SCHEDULER
```

Figure 4.5 Clock-driven scheduler.

4.2.1 Structured Clock-Driven Scheduling

Although the schedules in Example 4.3 are feasible, scheduling decision points are randomly scattered. In other words, there is no pattern in the time points at which new tasks are selected for execution. The idea with structured clock-driven scheduling approach is that scheduling decisions are made at periodical, rather than arbitrary, times. This way, a timer that expires periodically with a fixed length of duration can be used for decision times.

4.2.1.1 Frames

In a structured schedule, the scheduling decision times partition the time line into intervals called *frames*. Each frame is of a length f, which is called *frame size*. There is no preemption within a frame, because scheduling decisions are only made at the beginning of each frame. To ease the schedule construction, the phase of each periodic task is a nonnegative multiple of the frame size. In other words, the first instance of each periodic task is released at the beginning of some frame.

We just mentioned that there should be no preemption within a frame (otherwise, it is not a structured schedule). Since preemption involves context switches, it is desirable to avoid preemption. Therefore, when we select a frame size, it should be big enough to allow every task to complete its execution inside a frame. Assume that there are n periodic tasks in a system. This constraint can be expressed mathematically as

$$f \geq \max\{e_i, \, i = 1, \, 2, \, \ldots, \, n\} \tag{4.1}$$

To minimize the number of entries in a schedule table to save storage space, the chosen frame size should divide the major size. Otherwise, storing the schedule for one major cycle won't be sufficient, because the schedule won't simply repeat from one major cycle to another, and thus, a larger scheduling table is needed. This constraint can be formulated as

$$H \bmod f = 0 \tag{4.2}$$

Since H is a multiple of every task's period, if f can divide H, then f must also divide the period of at least one task in the system, that is,

$$p_i \bmod f = 0, \, \exists i \in \{1, 2, \ldots, n\}$$

Notice that this constraint is important, because storage is limited for most embedded systems.

When we choose the frame size, we should also need to ensure that it allows every task instance to meet its deadline. This constraint imposes that there is at least one full frame between the arrival of a task instance and its deadline. Because if a task instance arrives right after a frame starts, it won't be scheduled until the beginning of the following frame. However, a frame size can be up to

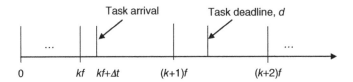

Figure 4.6 A task instance released right after a frame starts.

the execution time of a task. This may cause the instance to miss its deadline. Figure 4.6 illustrates this situation.

In Figure 4.6, a task instance arrives at time $kf + \Delta t$, where $\Delta t < f$. Its deadline d is between $(k + 1)f$ and $(k + 2)f$. This task instance will be scheduled for execution at the earliest $(k + 1)f$, the beginning of the next frame after it arrives. Before its deadline, it can only be executed for $d - (k + 1)f$ units of time, which is less than the frame size. If the execution time is close or equal to the frame size, then the task instance cannot finish execution before its deadline.

The constraint that there must be a full frame between the arrival of a task instance and its deadline is formulated as

$$d_i - (kf + \Delta t) \geq f + (f - \Delta t)$$

Because

$$d_i - (kf + \Delta t) = D_i$$

Therefore,

$$2f - \Delta t \leq D_i$$

Since the first instance of the task is released at the beginning of a frame, the minimum of Δt is the greatest common devisor (GCD) of p_i and f. The minimum of Δt corresponds to the worst case, that is, the task instance has the greatest chance to miss its deadline. Thus the third constraint can be written as

$$2f - \text{GCD}(p_i, f) \leq D_i, \quad i = 1, 2, \ldots, n. \tag{4.3}$$

Example 4.4 Cyclic Schedule
Consider a system of three periodic tasks:

$$T_1 = (4, 1), T_2 = (5, 1), T_3 = (10, 2).$$

We want to develop a cyclic scheduler for the system. First, we need to select a proper frame size.

According to the first constraint, we have $f \geq 2$.

The major cycle of the three tasks is $H = 20$. According to the second constraint, f should divide 20. Possible frame sizes are 2, 4, 5, and 10. We don't need to consider 1, because it violates the first constraint.

Now we need to test 2, 4, 5, and 10 with the third constraint $2f - GCD(p_i, f) \leq D_i$ for each task. Consider $f = 2$ first. For $T_1 = (4, 1)$,

$$2f - GCD(p_1, f) = 2 * 2 - GCD(4, 2) = 4 - 2 = 2$$

while $D_1 = 4$. Therefore, the constraint is satisfied by T_1.
For $T_2 = (5, 1)$,

$$2f - GCD(p_2, f) = 2 * 2 - GCD(5, 2) = 4 - 1 = 3,$$

while $D_2 = 5$. Therefore, the constraint is satisfied by T_2.
For $T_3 = (10, 2)$,

$$2f - GCD(p_3, f) = 2 * 2 - GCD(10, 2) = 4 - 2 = 2,$$

while $D_3 = 10$. Therefore, the constraint is satisfied by T_3.
Thus, the third constraint is satisfied by all tasks when $f = 2$. This also concludes that one choice of the frame size is 2.
Now let us examine $f = 4$. For $T_1 = (4, 1)$,

$$2f - GCD(p_1, f) = 2 * 4 - GCD(4, 4) = 8 - 4 = 4,$$

while $D_1 = 4$. Therefore, the constraint is satisfied by T_1.
For $T_2 = (5, 1)$,

$$2f - GCD(p_2, f) = 2 * 4 - GCD(5, 4) = 8 - 1 = 7,$$

while $D_2 = 5$. The inequality is not satisfied by T_2. There is no need to further test T_3.
Consider $f = 5$. For $T_1 = (4, 1)$,

$$2f - GCD(p_1, f) = 2 * 5 - GCD(4, 5) = 10 - 1 = 9,$$

while $D_1 = 4$. The inequality is not satisfied by T_1.
Consider $f = 10$. For $T_1 = (4, 1)$,

$$2f - GCD(p_1, f) = 2 * 10 - GCD(4, 10) = 20 - 2 = 18,$$

while $D_1 = 4$. The inequality is not satisfied by T_1.
Thus, the only feasible frame size for cyclic scheduling is 2. Figure 4.7 shows a feasible schedule with frame size 2.

The number of entries in the schedule table of a cyclic scheduler is equal to the number of frames in a major cycle. Each entry lists the schedule and the names of tasks scheduled to execute in the frame. Table 4.2 represents the

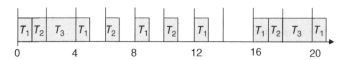

Figure 4.7 A cyclic schedule of frame size 2.

Table 4.2 The schedule table for the schedule shown in Figure 4.7.

Entry k	Schedule block $L(k)$
0	T_1, T_2
1	T_3
2	T_1
3	T_2
4	T_1
5	T_2
6	T_1
7	I
8	T_1, T_2
9	T_3

```
Input: Stored scheduling table L(k) for k= 0, 1, ..., F - 1.
Task CYCLIC_SCHEDULER:
        t := 0;  //current time
        k := 0; //frame number
        loop FOREVER
                accept clock interrupt;
                current Block := L(k);
                t++;
                k := t mod F;
                execute currentTask;
                sleep until the next clock interrupt;
        end loop
end CYCLIC_SCHEDULER
```

Figure 4.8 Cyclic scheduler.

schedule table of the schedule shown in Figure 4.7. The entry 7 is an I, which stands for *idle*.

In the cyclic scheduler shown in Figure 4.8, the schedule in a frame is called a *scheduling block*. Each entry in the schedule table is a block. The k-th block is denoted by $L(k)$, $k = 0, 1, 2, ..., F - 1$, where F is the number of frames in a major cycle. The clock periodically interrupts every f units of time.

4.2.1.2 Task Slicing
It is possible that for a set of periodic tasks, there is no feasible frame size that meets all the three constraints. This is typically because one or more tasks have large executions times. In that case, the first constraint may conflict with

the third constraint. One way to resolve the issue is to slice one or more big tasks (tasks with longer execution times) into several smaller ones. We use an example to illustrate the idea.

Example 4.5 *Task Slicing*
Given the following three periodic tasks:

$$T_1 = (4, 1), T_2 = (5, 1), T_3 = (10, 3),$$

we want to find a feasible frame size for a cyclic scheduler. The first constraint requires $f \geq 3$. The second constraint restricts the frame size to be 4, 5, or 10. Notice that the three tasks are similar to the ones in Example 4.4. The only difference is that we changed the execution of T_3 from 2 to 3 units of time. From the analysis in Example 4.4, we know that none of 4, 5, or 10 is a feasible frame size. Therefore, there is no frame size that meets all the three constraints. The reason is obvious: we have a big chunk task, T_3.

Assume that T_3 is preemptable. We slice T_3 to two smaller tasks:

$$T_{31} = (10, 2), T_{32} = (10, 1).$$

Now the problem becomes developing a cyclic scheduler for four tasks:

$$T_1 = (4, 1), T_2 = (5, 1), T_{31} = (10, 2), T_{32} = (10, 1).$$

It is easy to find out that one feasible frame size is 2.

Of course, we can also slice T_3 to three smaller tasks:

$$T_{31} = (10, 1), T_{32} = (10, 1), T_{33} = (10, 1).$$

This would mean more frequent preemption. Since preemption involves context switches, it is desirable to minimize the chance of preemption. Therefore, slicing T_3 into two smaller periodic tasks with the same period is a better option.

4.2.2 Scheduling Aperiodic Tasks

Aperiodic tasks are typically results of external events. They do not have hard deadlines. However, it is desirable to service aperiodic tasks as soon as possible in order to reduce the latency and improve the system's performance.

Aperiodic tasks are usually scheduled for execution at the background of periodic tasks. Therefore, aperiodic tasks are executed during the idle time slots of the processor. For example, in the schedule shown in Figure 4.7, the time intervals of [5, 6], [7, 8], [9, 10], [11, 12], and [13, 16] are idle. These idle time intervals are called *slacks*. They can be used to execute aperiodic tasks if they are waiting for execution.

Recall for periodic task instances, all that we need to ensure is that their deadlines are met; there is no benefit in completing their executions earlier. Therefore, we can try to delay the execution of periodic task instances as much as possible, as long as they can still meet their deadlines. This way, we can execute aperiodic tasks at the earliest possible times. This technique is called *slack stealing*.

Example 4.6   *Slack Stealing*

Figure 4.9a is a copy of the schedule in Example 4.4. Figure 4.9b shows the arrival of two aperiodic tasks. The first one arrives at 2 with 1.5 units of execution time. The second one arrives at 7 with 0.5 units of execution time. If we schedule them at the background of the schedule in Figure 4.9a, the first task will be completed at 10 and the second one will be completed at 12, as shown in Figure 4.9c. However, if we move the execution of the second instance of T2 to the interval of [9, 10] and the third instance of T2 to [11, 12], we are able to complete the first aperiodic task at 8 and the second aperiodic task at 11, as shown in Figure 4.9d.

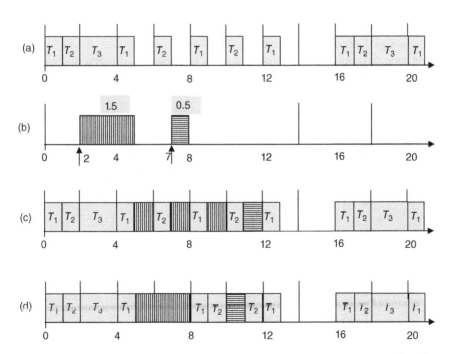

Figure 4.9 (a) Periodic task schedule. (b) Two aperiodic tasks. (c) Scheduling aperiodic tasks without slack stealing. (d) Scheduling aperiodic tasks with slack stealing.

4.2.3 Scheduling Sporadic Tasks

Sporadic tasks have hard deadlines, but their parameters (release time, execution time, and deadline) are unknown *a priori*, so there is no way for the scheduler to guarantee that they meet their deadlines. A common practice is that when a sporadic task is released with a known deadline, the scheduler will test to see if it can schedule the task such that its deadline can be met, given that all periodic tasks and maybe other sporadic tasks have already been scheduled. If the newly released sporadic task can pass the test, it will be scheduled for execution. Otherwise, the scheduler rejects the task and notifies the system so that it can take necessary recovery action.

Example 4.7 *Scheduling Sporadic Tasks*
Figure 4.10 shows a schedule of periodic tasks for two major cycles. The frame size of the scheduler is 4, and the major cycle is 20. Three sporadic tasks are released in the first major cycle. We want to test if they are schedulable or not.

At time 2, the first sporadic task S_1 is released with deadline 17 and execution time 3. The acceptance test is performed at 4, the beginning of the next frame. There are 4 units of slack time in total before 17, which is greater than the task's execution time. Therefore, S_1 passes the acceptance test. One unit of the S_1 is executed in the frame that starts at 4.

At time 5.5, the second sporadic task S_2 is released with deadline 19 and execution time 2. The acceptance test is performed at 8, the beginning of the following frame. There are 3 units of slack time in total from 8 to 19, which is greater than S_2's execution time. However, S_1 is not done yet; it still has 2 time units to finish. In addition, S_1 has an earlier deadline compared to S_2, so we should schedule S_1 first. After we reserve 2 units of execution time for S_1, there is only 1 time unit left for S_2 before its deadline, which is less than its execution time. Hence, the scheduler rejects S_2. S_1 completes its execution in the frame that starts at 8.

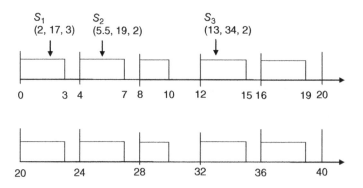

Figure 4.10 Scheduling sporadic tasks.

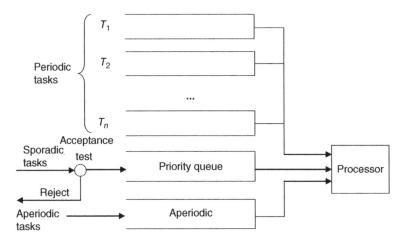

Figure 4.11 Scheduling of periodic, aperiodic, and sporadic tasks.

At time 13, the third sporadic task S_3 is released with deadline 34 and execution time 2. The acceptance test is performed at 16. There are 5 units of slack time in total between 16 and 34, which is greater than the task's execution time. Meanwhile, there is no other sporadic task waiting for execution. Therefore, S_3 passes the acceptance test. One unit of S_3 is executed in the frame that starts at 12, and the second unit executes in the frame that starts at 20.

In general, when periodic, aperiodic, and sporadic tasks are mixed together, the cyclic scheduler places all aperiodic tasks in one queue and all sporadic tasks in another queue. The sporadic task queue is a priority queue. The task with the earliest deadline is placed at the head of the queue. Of course, only sporadic tasks that pass the acceptance test are placed in the queue. Periodic tasks should be scheduled toward their deadlines as long as the deadlines are met, to allow aperiodic tasks to steal slacks and complete execution as early as possible. The scheduling scheme of all three types of tasks is illustrated in Figure 4.11.

4.3 Round-Robin Approach

The round-robin approach is a time sharing scheduling algorithm. In the round-robin scheduling, a time slice is assigned to each task in a circular order. Tasks are executed without priority. There is an FIFS (first-in-first-service) queue that stores all tasks in the *ready* state. Each time, the task at the head of the queue is removed from the queue and dispatched for execution. If the task is not finished within the assigned time slice, it is placed at the tail of the FIFS queue to wait for its next turn. Normally, the length of the time slice is short, so that the execution of every task appears to start almost immediately after it is ready. Since each time a task only gets a small portion executed, all tasks are assumed to be preemptable, and context switches occur frequently in this scheduling approach.

The round-robin scheduling algorithm is simple to implement. It is fair to all tasks in using the processor. The major drawback is that it delays the completion of all tasks and may cause tasks to miss their deadlines. It is not a good option for scheduling tasks with hard deadlines.

An improved version of round-robin is *weighted round-robin*. In the weighted round-robin approach, instead of assigning an equal share of the processor time to all ready tasks, different tasks may obtain different *weights*, the fractions of the processor time. By adjusting the weights, we can speed up or slow down the completion of a task.

4.4 Priority-Driven Scheduling Algorithms

In contrast to the clock-driven scheduling algorithms that schedule tasks at specific time points off-line, in a priority-driven scheduling algorithm, scheduling decisions are made when a new task (instance) is released or a task (instance) is completed. It is online scheduling, and decisions are made at runtime. Priority is assigned to each task. Priority assignment can be done statically or dynamically while the system is running. A scheduling algorithm that assigns priority to tasks statically is called a *static-priority* or *fixed-priority* algorithm, and an algorithm that assigns priority to tasks dynamically is said to be *dynamic-priority* algorithm.

Priority-driven scheduling is easy to implement. It does not require the information on the release times and execution times of tasks *a priori*. The parameters of each task become known to the online scheduler only after it is released. Online scheduling is the only option in a system whose future workload is unpredictable.

In this section, unless mentioned otherwise, we assume the following:

1) There are only periodic tasks in the system under consideration.
2) The relative deadline of each task is the same as its period.
3) All tasks are independent; there are no precedence constraints.
4) All tasks are preemptable, and the cost of context switches is negligible.
5) Only processing requirements are significant. Memory, I/O, and other resource requirements are negligible.

4.4.1 Fixed-Priority Algorithms

The most well-known fixed-priority algorithm is the *rate-monotonic* (RM) algorithm. The algorithm assigns priorities based on the period of tasks. Given two tasks $T_i = (p_i, e_i)$ and $T_j = (p_j, e_j)$, if $p_i < p_j$, then T_i has higher priority than T_j.

Scheduling periodic tasks based on the RM algorithm is relatively easy: when a new task instance is released, if the processor is idle, it executes the task; if the processor is executing another task, then the scheduler compares their priorities. If the new task's priority is higher, then it preempts the task in execution and executes on the processor. The preempted task is placed in the queue of ready tasks.

Example 4.8 *RM Scheduling*
Figure 4.12 shows an RM-based schedule of three periodic tasks:

$$T_1 = (4, 1), T_2 = (5, 1), T_3 = (10, 3).$$

Because $p_1 < p_2 < p_3$, T_1 has the highest priority and T_3 has the lowest priority. Whenever an instance from T_1 is released, it preempts whatever is running on the processor and gets executed immediately. Instances from T_2 run in the "background" of T_1. Task T_3, on the other hand, cannot execute when either T_1 or T_2 is unfinished.

Now let us examine the schedule and see if any task instance missed its deadline. We don't need to examine T_1, because whenever an instance is released, it gets executed on the processor immediately. For T_2, there are four instances released in the first major cycle, and they all get executed before their deadlines. For T_3, the first instance is completed at 7, which is before its deadline 10. The second instance is released at 10 and completed at 15, which is also before its deadline 20. Therefore, the three periodic tasks are schedulable based on RM.

Example 4.9 *Task missing deadline by RM*
Figure 4.13 shows that when we schedule three periodic tasks

$$T_1 = (4, 1), T_2 = (5, 2), T_3 = (10, 3.1).$$

according to the RM algorithm, the first instance of T_3 misses its deadline – it has remaining 0.1 units of execution time that is not completed by its deadline 10.

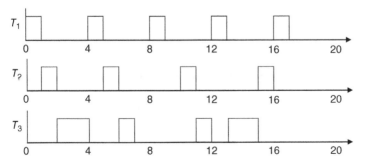

Figure 4.12 RM scheduling of three periodic tasks.

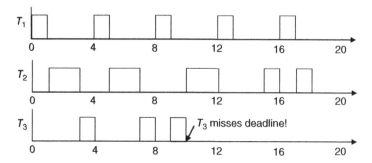

Figure 4.13 Example of task missing deadline in RM schedule.

4.4.1.1 Schedulability Test Based on Time Demand Analysis

We just saw two RM scheduling examples. In the first example, the RM produces a feasible schedule, but in the second example, it does not. You might wonder how to test whether a system can be scheduled by the RM algorithm or not.

Let us introduce a simple test first:

If the total utilization of a set of periodic tasks is not greater than

$$U_{RM}(n) = n(2^{1/n} - 1),$$

where n is the number of tasks, then the RM algorithm can schedule all the tasks to meet their deadlines.

Notice that this statement includes only a sufficient condition, but not necessary. That is, you may find a set of tasks with a total utilization greater than $n(2^{1/n} - 1)$, but they are schedulable by the RM algorithm.

For $n = 3$, we have $U_{RM}(3) = 0.78$. In Example 4.8, the total utilization is 0.75. Because $0.75 < 0.78$, the three tasks are schedulable as we proved in Figure 4.12. In Example 4.9, the total utilization is 0.96. Because $0.96 > 0.78$, we cannot draw any conclusion from the aforementioned statement. One thing we can claim based on this example is that the RM algorithm is not optimal, because, as we mentioned before, the SU of an optimal algorithm is 1.

Now let us discuss a more accurate approach to test whether a system can be scheduled by the RM algorithm. It is called *time demand analysis* (TDA). The TDA test is performed at *critical instants*. A critical instant of a task T_i is a time instant such that

- the instance in T_i released at the instant has the maximum response time of all instances in T_i, if the response time of every instance in T_i is equal to or less than the relative deadline D_i of T_i, and,
- the response time of the instance released at the instant is greater than D_i if the response time of some instances in T_i is greater than D_i.

Simply put, a critical instant is a time instant that has the worst combination of task instance release times. It is the instant that, when an instance in a task is released, all tasks of higher priority have an instance released at the same time, because in this case, the lower priority task instance has the longest delay before getting executed. Obviously, in a system that every task has phase 0, time 0 is a critical instant for every task. If a task can meet its deadline at a critical instant, then the task is schedulable by the RM algorithm.

Let us use a simple example to explain the idea of schedulability test using the TDA approach. Consider three periodic tasks T_1, T_2, and T_3, with T_1 having the highest priority and T_3 having the lowest priority. When an instance in T_1 is released, the scheduler interrupts whatever is running and places it for execution with no delay. Therefore, as long as the utilization of T_1 is not greater than 1, it is schedulable.

Consider the instance in T_2 released at time 0. It can complete its execution before its relative deadline p_2 if there is sufficient time for the instance in the interval $[0, p_2]$. Suppose that the instance completes at time t. The number of instances in T_1 released in the interval of $[0, t]$ is $\left\lceil \frac{t}{p_i} \right\rceil$. In order for the T_2 instance to finish at t, all those T_1 instances must be completed, and in addition, there must be e_2 units of the processor time available for executing the T_2 instance. Therefore, the total execution time or, in other words, the *demand for processor time* is

$$\left\lceil \frac{t}{p_1} \right\rceil e_1 + e_2$$

By the same token, in order for the first instance in T_3 to finish at t, all instances in T_1 and T_2 released in the interval of $[0, t]$ must be completed, and in addition, there must be e_3 units of the processor time available for executing the T_3 instance. Therefore, the demand for processor time is

$$\left\lceil \frac{t}{p_1} \right\rceil e_1 + \left\lceil \frac{t}{p_2} \right\rceil e_2 + e_3$$

For a general system with n tasks where $p_i < p_j$ for $j > i$, the demand for processor time to complete task T_i at time t is

$$w_i(t) = \sum_{j=1}^{i-1} \left\lceil \frac{t}{p_j} \right\rceil e_j + e_i \tag{4.4}$$

We call $w_i(t)$ the *time demand function* of T_i. On the other hand, the supply of processor time is t. T_i can meet its deadline at t if and only if the supply of processor time is greater than or equal to its time demand, that is,

$$w_i(t) \le t, \quad t \in [0, p_i] \tag{4.5}$$

Notice that $w_i(t)$ jumps when an instance in any higher priority task is released. Therefore, we test whether the inequality Eq. (4.5) is satisfied for some t that is a multiple of $p_1, p_2, \ldots$, or p_{i-1} over $0 \le t \le p_i$, or p_i. If and only if such a t exists, then is T_i schedulable; otherwise, it is not schedulable. If all tasks pass this test, then the system is schedulable by the RM algorithm.

Example 4.10 *Schedulability Test*

Consider the following four periodic tasks:

$$T_1 = (3, 1), T_2 = (4, 1), T_3 = (6, 1), T_4 = (12, 1).$$

We want to test their schedulability based on the TDA.

Task T_1 is schedulable, because $u_1 = 0.33 < 1$.

To test T_2, we first list the time instants when instances in T_1 are released in $[0, 4]$, and they are 0 and 3. If none of them satisfy the inequality Eq. (4.5), we shall further test the time instant 4, the deadline of the first instance in T_2. There is no need to test 0, because obviously it does not satisfy Eq. (4.5). At time 3,

$$\left\lceil \frac{t}{p_1} \right\rceil e_1 + e_2 = \left\lceil \frac{3}{3} \right\rceil 1 + 1 = 2$$

Inequality Eq. (4.5) is satisfied. Therefore, T_2 is schedulable.

The time instants when instances in T_1 and T_2 are released in $[0, 6]$ are 0, 3, 4 and 6. At time 3,

$$\left\lceil \frac{t}{p_1} \right\rceil e_1 + \left\lceil \frac{t}{p_2} \right\rceil e_2 + e_3 = \left\lceil \frac{3}{3} \right\rceil 1 + \left\lceil \frac{3}{4} \right\rceil 1 + 1 = 3$$

Inequality Eq. (4.5) is satisfied. Therefore, T_3 is schedulable.

The deadline of the first instance in T_4 is 12. The time instants when instances in T_1, T_2, and T_3 are released in $[0, 12]$ are 0, 3, 4, 6, 8, 9, and 12. At time 3,

$$\left\lceil \frac{t}{p_1} \right\rceil e_1 + \left\lceil \frac{t}{p_2} \right\rceil e_2 + \left\lceil \frac{t}{p_3} \right\rceil e_3 + e_4 = \left\lceil \frac{3}{3} \right\rceil 1 + \left\lceil \frac{3}{4} \right\rceil 1 + \left\lceil \frac{3}{6} \right\rceil 1 + 1 = 4 > 3$$

Inequality Eq. (4.5) is not satisfied. We have to continue the test with larger time points. At 4,

$$\left\lceil \frac{t}{p_1} \right\rceil e_1 + \left\lceil \frac{t}{p_2} \right\rceil e_2 + \left\lceil \frac{t}{p_3} \right\rceil e_3 + e_4 = \left\lceil \frac{4}{3} \right\rceil 1 + \left\lceil \frac{4}{4} \right\rceil 1 + \left\lceil \frac{4}{6} \right\rceil 1 + 1 = 5 > 4$$

Inequality Eq. (4.5) is still not satisfied. At 5,

$$\left\lceil \frac{t}{p_1} \right\rceil e_1 + \left\lceil \frac{t}{p_2} \right\rceil e_2 + \left\lceil \frac{t}{p_3} \right\rceil e_3 + e_4 = \left\lceil \frac{5}{3} \right\rceil 1 + \left\lceil \frac{5}{4} \right\rceil 1 + \left\lceil \frac{5}{6} \right\rceil 1 + 1 = 6 > 5$$

Inequality Eq. (4.5) is still not satisfied. At 6,

$$\left\lceil \frac{t}{p_1} \right\rceil e_1 + \left\lceil \frac{t}{p_2} \right\rceil e_2 + \left\lceil \frac{t}{p_3} \right\rceil e_3 + e_4 = \left\lceil \frac{6}{3} \right\rceil 1 + \left\lceil \frac{6}{4} \right\rceil 1 + \left\lceil \frac{6}{6} \right\rceil 1 + 1 = 6$$

Now the inequality is satisfied. Therefore, T_4 is schedulable. This means that all tasks in the system pass the schedulability test.

Example 4.11 *System Fails TDA*

Example 4.9 shows that the following three periodic tasks

$$T_1 = (4, 1), T_2 = (5, 2), T_3 = (10, 3.1)$$

cannot be scheduled by the RM algorithm. Now let us prove it with TDA.

For task T_2, the two time instants to test its schedulability are 4 and 5. At time 4,

$$\left\lceil \frac{t}{p_1} \right\rceil e_1 + e_2 = \left\lceil \frac{4}{4} \right\rceil 1 + 2 = 3 < 4$$

Inequality Eq. (4.5) is satisfied. Therefore, T_2 is schedulable.

The schedulability of T_3 can be tested at time instants 4, 5, 8, and 10. At time 4,

$$\left\lceil \frac{t}{p_1} \right\rceil e_1 + \left\lceil \frac{t}{p_2} \right\rceil e_2 + e_3 = \left\lceil \frac{4}{4} \right\rceil 1 + \left\lceil \frac{4}{5} \right\rceil 2 + 3.1 = 6.1 > 4$$

Inequality Eq. (4.5) is not satisfied. We have to continue the test with the next time point. At 5,

$$\left\lceil \frac{t}{p_1} \right\rceil e_1 + \left\lceil \frac{t}{p_2} \right\rceil e_2 + e_3 = \left\lceil \frac{5}{4} \right\rceil 1 + \left\lceil \frac{5}{5} \right\rceil 2 + 3.1 = 7.1 > 5$$

Inequality Eq. (4.5) is not satisfied. We test it with time 8:

$$\left\lceil \frac{t}{p_1} \right\rceil e_1 + \left\lceil \frac{t}{p_2} \right\rceil e_2 + e_3 = \left\lceil \frac{8}{4} \right\rceil 1 + \left\lceil \frac{8}{5} \right\rceil 2 + 3.1 = 9.1 > 8$$

It still does not work. The last time point to test is time 10. At 10,

$$\left\lceil \frac{t}{p_1} \right\rceil e_1 + \left\lceil \frac{t}{p_2} \right\rceil e_2 + e_3 = \left\lceil \frac{10}{4} \right\rceil 1 + \left\lceil \frac{10}{5} \right\rceil 2 + 3.1 = 10.1 > 10$$

The inequality is still not satisfied. We have tested all time points. Therefore, T_3 is not schedulable. This is consistent with what is revealed in Example 4.9.

If we plot $w_i(t)$ over t, then T_i is schedulable if and only if some part of the plot of $w_i(t)$ falls on or below the $w_i(t) = t$ line before or at its relative deadline p_i. Figure 4.14 shows the plots of $w_i(t)$ for T_1, T_2, and T_3. The dotted line indicates the locus of $w_i(t) = t$. As you can see, $w_1(t)$ has a segment below the dotted line before its deadline 4, so does $w_2(t)$ before 5. However, $w_3(t)$ is all above the dotted line from 0 to 10.

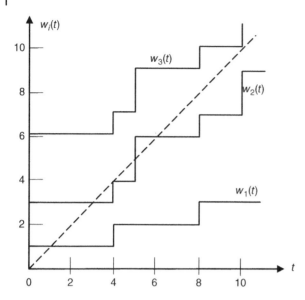

Figure 4.14 Time demand function for tasks in Example 4.11.

4.4.1.2 Deadline-Monotonic Algorithm

Besides the RM algorithm, another well-known fixed-priority scheduling algorithm is *deadline-monotonic* (DM), which assigns priorities to tasks based on their relative deadlines. A task with shorter relative deadline has higher priority than those with longer relative deadlines. For example, for a system with three tasks $T_1 = (50, 50, 25, 100)$, $T_2 = (0, 62.5, 10, 20)$, and $T_3 = (0, 125, 25, 50)$, T_2 has the highest priority while T_1 has the lowest priority. If in a system every task's relative deadline is equal to its period, then RM and DM will produce the same schedule.

4.4.2 Dynamic-Priority Algorithms

Fixed-priority scheduling algorithms assign a fixed priority to all instances in a task. It is not the case in a *dynamic-priority* algorithm, in which different instances in a task may be assigned with different priorities. The most widely used dynamic-priority scheduling algorithm is the *earliest-deadline-first* (EDF) algorithm.

4.4.2.1 Earliest-Deadline-First (EDF) Algorithm

A scheduler following the EDF algorithm always schedules the task whose absolute deadline is the earliest for execution. A task's priority is not fixed. It is decided at runtime based on how close it is to its absolute deadline.

Next, we first use three one-shot tasks to show how to apply the EDF algorithm in task scheduling. After that, we show that the three periodic tasks in

Example 4.9 are schedulable by the EDF algorithm (they are not schedulable by the RM algorithm).

Example 4.12 *EDF Scheduling of One-Shot Tasks*

Consider the four one-shot hard deadline tasks listed in Table 4.3. Assume that they are all independent and preemptable. At time 0, T_1 is released. Since it is the only task in the system, it is scheduled for execution. At time 1, T_2 is released and its deadline is at 4, which is earlier than the deadline of T_1. Therefore, T_2 has higher priority than T_1. Although T_1 still has one unit that is not finished, it is preempted by T_2. So, at time 1, T_2 executes. Because its execution time is 1, so it is done at time 2. Then T_1 resumes its execution at time 2, and it is completed at 3. At 3, T_3 is released and its deadline is at 10. Now T_3 is the only task in the ready queue, and so it is scheduled for execution. At time 5, T_4 is released and its deadline is at 8, earlier than the deadline of T_3. So, T_4 preempts T_3 and executes. At 7, T_4 is done and T_3 resumes execution. T_3 is completed at 8. The schedule is shown in Figure 4.15.

Example 4.13 *EDF Scheduling of Periodic Tasks*

In Examples 4.9 and 4.11, we showed that the following three independent, preemptable periodic tasks are not schedulable by the RM scheduling algorithm:

$$T_1 = (4, 1), T_2 = (5, 2), T_3 = (10, 3.1)$$

Now let us examine if we can schedule them by the EDF algorithm.

At 0, the first instance in each task is released. Because the instance in T_1 has the earliest deadline, it executes. It is done at 1.

At 1, the instance in T_2 has higher priority than the instance in T_3 and executes. It is done at 3.

At 3, T_3 is the only task waiting to run, and so it executes.

Table 4.3 The aperiodic tasks in Example 4.12.

Task	Release time	Execution time	Deadline
T_1	0	2	6
T_2	1	1	4
T_3	3	3	10
T_4	5	2	8

Figure 4.15 EDF schedule of three one-shot tasks.

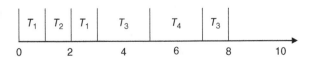

At 4, the second instance in T_1 is released. Its deadline is at 8, earlier than the deadline of the T_3 instance, which is at 10. So, the T_1 instance preempts T_3 and executes. It is done at 5.

At 5, the second instance in T_2 is released. Its deadline is at 10, same as the deadline of the T_3 instance. Since T_3 arrives earlier, T_3 executes and is completed at 7.1, which means that the first instance of T_3 meets its deadline!

At 7.1, T_2 is the only task waiting to run, and so it executes.

At 8, the third instance in T_1 arrives. Its deadline is at 12, later than the deadline of the T_2 instance in execution, which still has 1.1 units to complete. So, T_2 continues to run and completes at 9.1.

At 9.1, T_1 is the only task waiting to run, and so it executes.

At 10, an instance in T_2 is released with deadline 15, and meanwhile, an instance in T_3 is released with deadline at 20. So, the T_1 instance continues to run. Both T_2 and T_3 are waiting.

At 10.1, T_1 is completed. T_2 has higher priority than T_3. The instance in T_2 executes. T_3 is waiting.

At 12, an instance in T_1 is released with deadline 16. The T_2 instance continues to run. T_1 and T_3 are waiting.

At 12.1, T_2 is completed. T_1 has higher priority than T_3. The instance in T_1 executes. T_3 is waiting.

At 13.1, T_1 is completed. The instance in T_3 executes. No task is waiting.

At 15, an instance in T_2 is released with deadline at 20. It has the same priority as the T_3 instance in execution. T_3 continues to run. T_2 is waiting.

At 16, an instance in T_1 is released with deadline at 20. It has the same priority as T_2 and T_3. T_3 continues to run. T_1 and T_2 are waiting.

At 16.2, T_3 is completed. T_1 and T_2 have the same deadline, but T_2 arrives earlier. The instance in T_2 executes. T_1 is waiting.

At 18.2, T_2 is completed. T_1 executes. No task is waiting.

At 19.2, T_1 is completed. No task is waiting.

We have completed the scheduling for the first major cycle of the three tasks. Every task instance meets its deadline. Therefore, the tasks are schedulable by the EDF algorithm. The schedule is plotted in Figure 4.16.

4.4.2.2 Optimality of EDF

Example 4.13 shows that EDF produces feasible schedule for the three periodic tasks that cannot be scheduled by RM. In fact, EDF is an optimal uniprocessor scheduling algorithm. In other words, as long as a set of tasks has feasible schedules, EDF can produce a feasible schedule.

We prove that EDF is optimal by showing that any feasible schedule can be systematically transformed to an EDF schedule. Assume that an arbitrary schedule S meets all deadlines. If S is not an EDF schedule, there must be a situation as illustrated in Figure 4.17a, that is, a task with a later deadline is scheduled to run before a task with an earlier deadline. Assume before that the schedule is EDF.

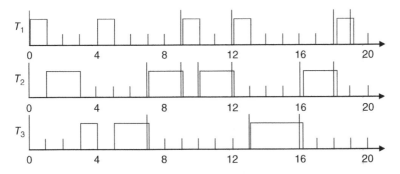

Figure 4.16 EDF schedule of periodic tasks that cannot be scheduled by the RM algorithm.

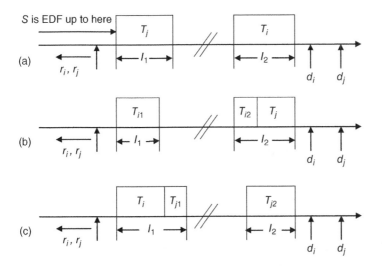

Figure 4.17 Proof of optimality of EDF.

We consider three cases.

Case 1: Interval I_1 is shorter than I_2. In this case, we partition T_i into two subtasks T_{i1} and T_{i2}, where T_{i1} is the first portion of T_i and its execution time is equal to the length of I_1. We place T_{i1} in I_1, move T_{i2} to the beginning of I_2, and place I_j right after I_{i2}, as shown in Figure 4.17b.

Case 2: Interval I_1 is longer than I_2. In this case, we partition T_j into two subtasks T_{j1} and T_{j2}, where T_{j1} is the first portion of T_j and its execution time is equal to the difference in the lengths of I_1 and I_2. We place T_i and T_{j1} in I_1 and place T_{j2} in I_2, as shown in Figure 4.17c.

Case 3: Intervals I_1 and I_2 are equally long. In this case, we simply switch the two tasks.

After the adjustment in either case, the two tasks tend to be in the EDF order. Notice that the adjustment does not affect the rest of the schedule. Therefore, if we repeat this process for any two tasks that are out of order, we turn the schedule to an EDF schedule. This proves that any feasible schedule can be systematically transformed to an EDF schedule. Thus, EDF is optimal.

How do we test the schedulability of a set of tasks by the EDF algorithm? We have an important statement as follows:

If all tasks in a system are periodic and have relative deadlines equal to their respective periods and the total utilization of the tasks is not greater than 1, then the system is schedulable on a single processor by the EDF algorithm.

The necessary condition is obvious, because the utilization of a single processor cannot exceed 1. We focus on the proof of the sufficient condition. We prove it by showing that if in an EDF schedule, some tasks of a system fail to meet their deadlines, then the total utilization of the system must be greater than 1.

Without loss of generality, assume that the system starts execution at 0, and the event that the first task, say T_i, misses its deadline occurs at time t.

Assume that the processor never idles prior to t. Consider two cases: (1) the current period of every task starts at or after r_i, the release time of the instance in T_i that misses its deadline at t, and (2) the current periods of some tasks begin before r_i. The two cases are illustrated in Figure 4.18.

Case (1): This case is illustrated in Figure 4.18a. Each tick on the time line of a task shows the release time of some instance in the task. The fact that T_i misses its deadline at t indicates two things:

a) Any current task instance whose deadline is after t is not given any processor time to execute before t.

b) The total processor time required to complete the T_i instance and all other task instances with deadlines at or before t exceeds the total supply of processor time, t.

In other words, we have

$$t < \frac{(t - \phi_i)e_i}{p_i} + \sum_{k \neq i} \left\lfloor \frac{t - \phi_k}{p_k} \right\rfloor e_k \qquad (4.6)$$

The first term on the right-hand side of the inequality is the time required to execute all instances in T_i with deadlines before t. Each term in the sum gives the total amount of time before t required to complete instances that are in a task T_k other than T_i and have deadlines at or before t. Notice that the flooring function excludes the task instances whose deadlines are after t. Since $\phi_k \geq 0$ and $e_k/p_k = u_k$ for all k, and $\lfloor x \rfloor \leq x$ for any $x \geq 0$, we have

$$\frac{(t - \phi_i)e_i}{p_i} + \sum_{k \neq i} \left\lfloor \frac{t - \phi_i}{p_k} \right\rfloor e_k \leq t\frac{e_i}{p_i} + t\sum_{k \neq i} \frac{e_k}{p_k} = t\sum_{k=1}^{n} u_k = tU$$

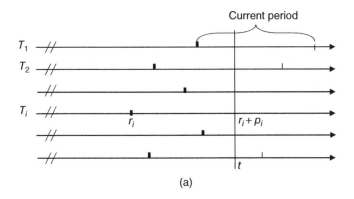

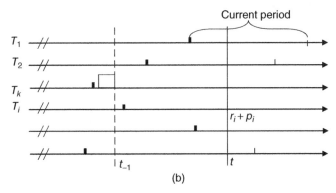

Figure 4.18 Infeasible EDF schedules.

Combining this inequality with the previous inequality, we have $U > 1$.

Case (2): This case is illustrated in Figure 4.16b. Denote by T' the set of tasks whose current period of instances are released before r_i and have deadlines after t. It is possible that processor time before r_i was given to the current instances of some tasks in T'. T_k is such a task in the figure. Let t_{-1} be the last point in time before r_i when some current instance in T' is executed, as illustrated in the figure, where an instance of T_k ends its execution at t_{-1}. Then, during the interval $[t_{-1}, t]$, no instances in tasks in $T - T'$ whose deadlines are after t are given processor time. Denote by ϕ'_j the release time of the first instance of task T_j in $T - T'$ after t_{-1}.

Because the current instance in T_i misses its deadline, we must have

$$t - t_{-1} < \frac{(t - t_{-1} - \phi'_i)e_i}{p_i} + \sum_{T_k \in T - T'} \left\lfloor \frac{t - t_{-1} - \phi'_k}{p_k} \right\rfloor e_k$$

This inequality is essentially the same as the one in Eq. (4.6) except that t is replaced with $t - t_{-1}$ and ϕ_k is replaced with ϕ'_k. This implies that $U > 1$.

Now let us consider the case that processor idles for some time prior to t. Let t_{-2} be the last time instant when the processor idles. In other words, from t_{-2} to t, the processor never idles. For the same reason, the inequality in Eq. (4.6) is true. Therefore, we have $U > 1$.

The total utilization of the three periodic tasks in Example 4.13 is 0.96. Therefore, they are schedulable by the EDF algorithm.

4.4.3 Priority-Driven Scheduling of Aperiodic and Sporadic Tasks

We have discussed the priority-driven scheduling of periodic tasks. In this section, we introduce algorithms for scheduling aperiodic tasks and sporadic tasks.

4.4.3.1 Scheduling of Aperiodic Tasks

Aperiodic tasks have either soft deadlines or no deadlines. The simplest algorithm for aperiodic task scheduling is to execute them at the background of periodic tasks, that is, at the slack times of the schedule of periodic tasks. The algorithm assigns the lowest priority to aperiodic tasks. Therefore, the schedulability of periodic tasks won't be affected, but there is no guarantee of the response time of aperiodic tasks.

We can apply the slack stealing technique to improve the scheduler's performance regarding the response to aperiodic tasks. The theory behind slack stealing is that for periodic tasks, the scheduling target is to meet their deadlines; there is no benefit to complete their execution earlier. Therefore, we can delay their execution as much as possible, as long as their deadlines are met. This way, we can execute aperiodic tasks at the earliest possible times. Slack stealing can be easily implemented in a clock-driven scheduling system, because a clock-driven schedule is computed off-line, and we know exactly when the processor is available for aperiodic tasks, how much the slack time is in each frame, and the maximum that a schedule can be move around. However, implementing slack stealing in a priority-driven scheduling system is much more complicated, because scheduling decisions are made at runtime.

A more popular approach to scheduling aperiodic tasks is *polling*. In this approach, a *polling server* or *poller* is introduced as a periodic task $T_s = (p_s, e_s)$ to the system, where p_s is the polling period and e_s is its execution time. The scheduler treats the polling server as one more periodic task and assigns a relative priority to it based on its polling period. When the poller executes, it examines the aperiodic task queue. If it is backlogged, the poller executes the task at the head of the queue. The poller stops running when it has already executed for e_s units of time or the aperiodic task that executes is finished, whichever comes first, and then waits for next cycle. However, if at the beginning of a polling period, no aperiodic tasks are available for execution, the poller suspends itself immediately. It won't be ready for execution until the next polling period. According to this polling policy, if an aperiodic task arrives right after the beginning of a polling period, it has to wait until the beginning of the next period to be considered for execution.

A *deferrable server* $T_{ds} = (p_s, e_s)$ is a bandwidth-preserving polling server that preserves the execution time or *budget* of the server until it is used up by executing aperiodic tasks in the current period or expires at the end of the period. This way, if an aperiodic task arrives at an empty aperiodic task queue after the beginning of a polling period, it still can get executed within the period if the server's priority permits. The idea of deferrable server is simple, but it reduces the delay of execution of aperiodic tasks.

Example 4.14 *Scheduling of Aperiodic Tasks*
Consider a system of two periodic tasks:

$$T_1 = (3, 1), T_2 = (10, 3).$$

They are scheduled by RM. An aperiodic task A with an execution time of 1.3 is released at 0.2. If we schedule A at the background of T_1 and T_2, its execution is completed at 7.3, as shown in Figure 4.19a.

If A is scheduled with the simple polling server $T_s = (2.5, 0.5)$, its execution is finished at 7.8, as shown in Figure 4.19b. Notice that since the period of T_s is shorter than T_1 and T_2, T_s has the highest priority. In addition, because A arrives after the first period begins, it is not executed until time 2.5, the beginning of the next period.

If A is scheduled with a deferrable server $T_s = (2.5, 0.5)$, its execution is finished at 5.5 as shown in Figure 4.19c. In this case, A executes immediately after it is released at 0.2. It is executed for 0.5 units of time in the first period, another 0.5 units of time in the second period. The last 0.3 is completed in the third period. The first instance of T_1 is preempted by the server at 0.2. It is resumed for execution at 0.7, the instant when A uses up its budget in the first period, and completes at 1.5.

Now let us examine the impact of a deferrable server to the schedulability of periodic tasks. Assume that the server has the highest priority. The worst impact occurs when an aperiodic task with execution time no less than $2e_s$ arrives at $mp_s - e_s$, where m is a nonnegative integer. In this case, the server executes for $2e_s$ units of time consecutively and causes the longest possible delay to periodic tasks. Therefore, in a system of periodic tasks scheduled by the RM algorithm, if aperiodic tasks are scheduled by a deferrable server that has the

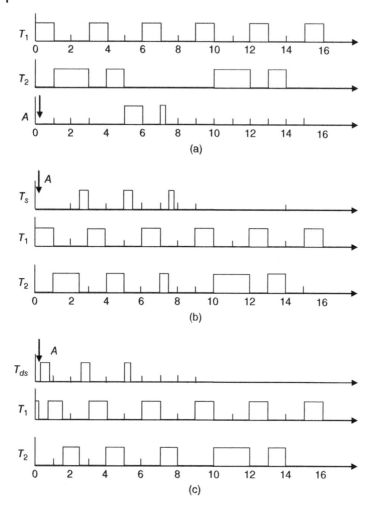

Figure 4.19 Aperiodic task scheduling. (a) Aperiodic task scheduled at the background of periodic tasks. (b) Aperiodic task scheduled with a simple polling server. (c) Aperiodic task scheduled with a deferrable server.

shortest period, a critical instant of every periodic task T_i occurs at time t_0 when all the following are true:

1) One of its instances is released at t_0.
2) An instance in every higher priority task is released at t_0.
3) The budget of the server at t_0 is e_s, one aperiodic task is released at t_0, the next period of the server starts at $t_0 + e_s$, and the server keeps running in the interval $[t_0, t_0 + 2e_s]$.

Hence, the time demand function of T_i is

$$w_i(t) = \left\lceil \frac{t - e_s}{p_s} \right\rceil e_s + e_s + \sum_{j=1}^{i-1} \left\lceil \frac{t}{p_j} \right\rceil e_j + e_i, \quad 0 < t \le p_i \qquad (4.7)$$

On the right-hand side of Eq. (4.7), the first two terms count the maximum processor time demand by the server at time t. Each term in the sum is the time demand by each higher priority task.

From the schedulability analysis we presented for periodic tasks, we know that if we can find a t in the interval $[0, p_i]$ such that $w_i(t) \le t$, then T_i is schedulable. This schedulability test can help us in selecting the two server parameters p_s and e_s. First of all, we don't want to introduce any deferrable server that will cause a periodic task to miss its deadline. From that end, the server should have a relatively big p_s and small e_s. However, in order for the system to respond quickly to aperiodic tasks, the utilization of the server cannot be too small. The schedulability analysis can help us figure out an acceptable server for both periodic tasks and aperiodic tasks.

4.4.3.2 Scheduling of Sporadic Tasks

Sporadic tasks are released irregularly. Although a sporadic task does not have a period, there must be a minimum interarrival time between any two consecutive instances of the task. Otherwise, it is hard for a scheduler to guarantee that all its instances meet their deadlines. One way of handling sporadic tasks is to treat every sporadic task as a periodic task with a period equal to its minimum interarrival time. Another way is to introduce a deferrable server, the way we handle aperiodic tasks. Since sporadic tasks have hard deadlines, an acceptance test should be performed before a sporadic task is scheduled for execution. If it is not schedulable, it should be rejected.

4.4.4 Practical Factors

The scheduling algorithms we have presented so far are based on assumptions that every task is preemptable at any time, once a task instance is released, it never suspends itself; hence, it is ready for execution until it completes, and context switch overhead is negligible compared to task execution time. However, in real-world applications, these assumptions are not always valid. In this section, we discuss how to test the schedulability of a system where these practical factors cannot be ignored.

4.4.4.1 Nonpreemptivity

A task or a particular portion of a task may be nonpreemptable. For example, when a task is running in its critical section, making it nonpreemptable is one way to avoid unbounded priority inversion. This will be discussed in detail in the next chapter. Tasks are also made not preemptable if preemption is too costly.

For example, consider a task that packages and sends control signal to an external device. If the portion of sending a control signal is preempted, it has to start over.

Denote by θ_i the largest nonpreemptable portion of T_i. A task is said to be *blocked* if it is prevented from being executed by lower priority job, a situation of *priority inversion*. When we test the schedulability of a task T_i, we must consider the nonpreemptable portions of lower priority tasks as well as the execution times of higher priority tasks.

The *blocking time* $b_i(np)$ of a task T_i is the longest time by which any instance of T_i can be blocked by lower priority tasks. In a fixed-priority system where tasks are indexed in the decreasing order of priorities, $b_i(np)$ is given by

$$b_i(np) = \max_{i+1 \leq k \leq n} \theta_k \tag{4.8}$$

4.4.4.2 Self-Suspension

Self-suspension of a task occurs when the task waits for an external operation, such as Remote Procedure Call (RPC) and I/O operation, to complete on another processor. When a task self-suspends, it loses the processor and is placed in a blocked queue by the scheduler. Of course, when the blocked task tries to reacquire the processor, it may be blocked by the tasks in their nonpreemptable portions.

The suspension of a task may delay the execution of lower priority tasks. Figure 4.20 illustrates such a case. Two tasks shown in the figure are

$$T_1 = (5, 2), T_2 = (2, 6, 2.1, 6).$$

The first instance of T_1 suspends itself for 3 units of time immediately after it starts execution, which causes the first instance of T_2 to miss its deadline at time 8.

Assume that we know the maximum length of external operation, that is, the duration of self-suspension is bounded. Denote by ρ_i the maximum self-suspension duration of T_i. Denote by $b_i(ss)$ the blocking time of a task T_i

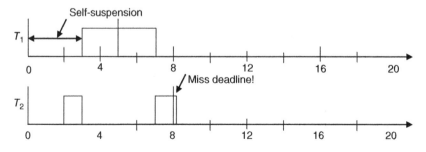

Figure 4.20 Illustration of impact of task self-suspension.

due to self-suspension. $b_i(ss)$ is given by

$$b_i(ss) = \rho_i + \sum_{k=1}^{i-1} \min\{e_k, \rho_k\} \tag{4.9}$$

4.4.4.3 Context Switches

Each task has a context, which is the data indicating its execution status and stored in the task control block (TCB), a data structure that contains all the information that is pertinent to the execution of the task. When a scheduler switches a task out of the CPU, its context has to be stored; when the task is selected to run again, its context is restored so that the task can be executed from the last interrupted point. In a fixed-priority system, each task preempts at most one lower priority task if there is no self-suspension. Hence, each task instance suffers one context switch when it starts execution and another context switch when it completes. Therefore, we can add the context switch time twice to the execution time of each task. Let CS be the maximum amount of time that the system spends on a single context switch, then when we analyze the schedulability of a system, we need to add $2CS$ to the execution time of each task if there is no self-suspension. For a task that can self-suspend up to k_i times, we add $2(k_i + 1)CS$ to its execution time.

4.4.4.4 Schedulability Test

In a fixed-priority system where a task T_i suspends k_i times, its total blocking time b_i is given by

$$b_i = b_i(ss) + (k_i + 1)b_i(np) \tag{4.10}$$

After adding the context switch time to each task's execution time, the time demand function $w_i(t)$ of the task T_i is modified as follows:

$$w_i(t) = \sum_{j=1}^{i-1} \left\lceil \frac{t}{p_j} \right\rceil e_j + e_i + b_i \tag{4.11}$$

We then analyze the schedulability of tasks based on this expanded time demand function.

Example 4.15 *Schedulability Test*

Consider four periodic tasks as follows:

$$T_1 = (3, 1), T_2 = (4, 1), T_3 = (6, 1), T_4 = (12, 1).$$

We showed in Example 4.10 that all these tasks are schedulable according to the RM algorithm. Now we assume the following:

- T_3 has a 0.2 time units of nonpreemptable portion, that is, $\theta_3 = 0.2$. All other tasks are preemptable at any time.

- T_1 may suspend itself at most once for 0.2 units of time. That is, $k_1 = 1$ and $p_1 = 0.2$. All other tasks have no self-suspension.
- Overhead of context switches is not negligible; $CS = 0.1$.

Let us find out if all the tasks are still schedulable.

According to Eqs. (4.8) and (4.9), we calculate $b_i(np)$ and $b_i(ss)$ for each task as follows:

$$b_1(np) = b_2(np) = q_3 = 0.2;$$

$$b_3(np) = b_4(np) = 0;$$

$$b_3(np) = b_4(np) = 0;$$

$$b_1(ss) = 0;$$

$$b_2(ss) = b_3(ss) = b_4(ss) = \min\{e_1, r_1\} = 0.2.$$

Then it results from Eq. (4.10) that

$$b_1 = b_1(ss) + (k_1 + 1)b_1(np) = 0.4;$$

$$b_2 = b_2(ss) + (k_2 + 1)b_2(np) = 0.4;$$

$$b_3 = b_3(ss) + (k_3 + 1)b_3(np) = 0.2;$$

$$b_4 = b_4(ss) + (k_4 + 1)b_4(np) = 0.2.$$

The execution time with context switch time included for each task is

$$e_1 = 1 + 2(k_1 + 1)CS = 1.4;$$

$$e_2 = e_3 = e_4 = 1 + 2CS = 1.2.$$

Now we can test the schedulability of each task based on the modified time demand function. For the task T_1,

$$w_1(t) = e_1 + b_1 = 1.4 + 0.4 = 1.8$$

At $t = 3$ we have $t > w_1(t)$. Therefore, T_1 is schedulable.
For the task T_2,

$$w_2(t) = \left\lceil \frac{t}{p_1} \right\rceil e_1 + e_2 + b_2 = \left\lceil \frac{t}{3} \right\rceil 1.4 + 1.2 + 0.4 = \left\lceil \frac{t}{3} \right\rceil 1.4 + 1.6$$

At $t = 3$ the inequality $t \geq w_2(t)$ is satisfied. Therefore, T_2 is schedulable.
For the task T_3,

$$w_3(t) = \left\lceil \frac{t}{p_1} \right\rceil e_1 + \left\lceil \frac{t}{p_2} \right\rceil e_2 + e_3 + b_3 = \left\lceil \frac{t}{3} \right\rceil 1.4 + \left\lceil \frac{t}{4} \right\rceil 1.2 + 1.4$$

The time instants to test the schedulability of T_3 are 3, 4, and 6. However, none of them satisfies $t \geq w_3(t)$. Therefore, T_3 is not schedulable.

4.5   Task Assignment

The scheduling algorithms presented in the previous sections are all uniprocessor-based. In reality, many real-time embedded systems are running on multiple processors because a single processor is not able to handle all tasks. As mentioned earlier in this chapter, a multiprocessor scheduling problem is often dealt with through uniprocessor scheduling by assigning tasks in a system to each processor first. In this section, we introduce some well-known task assignment approaches.

4.5.1   Bin-Packing Algorithms

In general, task execution times, communication costs between tasks when they are assigned to different processors, and placement of resources are among the most important factors that we need to consider when we perform task assignment. Bin-packing algorithms are utilization-based and do not take communication costs into consideration. In some real-time applications, tasks assigned to different processors exchange information through shared memory. Their communication costs are negligible. Bin-packing algorithms are good for the task assignment of this kind of systems. Besides, at the early stage of system design, we may want to ignore communication and resource access costs despite the fact that they may be significant. We estimate the execution time of each task based on its complexity and determine its utilization roughly, and then based on each task's utilization, we carry out task assignment.

The first step of task assignment is to decide the number of processors that are needed. This can be done roughly based on the total task utilization. For example, consider a system that contains n independent and preemptable periodic tasks whose relative deadlines are equal to their respective periods, and these tasks are to be scheduled by the EDF algorithm. If we know the utilization of each task, then we can easily compute the total utilization U of the system. Because the SU of EDF is 1, we round up U to the nearest integer, say k. Then k is the minimum number of processors that we need for the system. To assign the tasks to each processor, we can partition the set of tasks into k groups and assign all tasks in each group to the same processor. All that we need to make sure is that the total utilization of tasks in each group is not greater than 1. Of course, it may not be a good idea to attempt to fully utilize the processor time in periodic tasks. To reserve certain amount of processor time for the execution of aperiodic tasks and sporadic tasks, we can set the threshold to utilization of each processor to some value $U' < 1$.

It is straightforward to formulate this kind of task assignment problems as simple bin-packing problems. In a bin-packing problem, objects or items of different volumes must be packed into a finite number of bins or containers of the same volume in a way that minimizes the number of bins used. In the context of task assignment, objects are tasks, the number of bins is the number of processors, and the volume of each bin is U'. For n tasks with utilizations $u_1, u_2, \ldots, u_n$, we want to find the required number of processors N_p and a N_p-partition of the task set such that

$$\min N_p = \sum_{i=1}^{n} y_i \tag{4.12a}$$

$$\text{subject to } N_p \geq 1 \tag{4.12b}$$

$$\sum_{j=1}^{n} u_j x_{ij} \leq U' y_i, \quad i \in \{1, 2, \ldots, n\} \tag{4.12c}$$

$$\sum_{i=1}^{n} x_{ij} = 1, \quad j \in \{1, 2, \ldots, n\} \tag{4.12d}$$

$$y_i \in \{0, 1\}, \quad i \in \{1, 2, \ldots, n\} \tag{4.12e}$$

$$x_{ij} \in \{0, 1\}, \quad i \in \{1, 2, \ldots, n\}, \quad j \in \{1, 2, \ldots, n\} \tag{4.12f}$$

This is a integer linear programming (ILP) formulation of the problem. Equation (4.12c) states that the cap of the utilization of each processor is not greater than U'. Equation (4.12d) together with Eq. (4.12f) states that each task can only be assigned to one processor. Equation (4.12e) states that processor i may or may not be used. Notice that it is assumed that the maximum number of processors that are needed is n, the number of tasks.

4.5.1.1 First-Fit Algorithm

The bin-packing problem is NP-complete, but there are many simple heuristic algorithms to solve the problem. The first-fit algorithm is a very straightforward greedy approximation algorithm. It processes the items in arbitrary order. For each item, it attempts to place the item in the first bin that can accommodate the item. If no bin is found, it opens a new bin and places the item in the new bin.

According to this algorithm, given a set of tasks $T_1, T_2, \ldots, T_n$ and the maximum utilization of each processor U', we follow the following simple rules to perform the task assignment:

- Tasks are assigned one by one in an arbitrary order.
- Assign task T_1 to processor P_1.
- Assign T_i to P_k if
  - After the assignment, total utilization of P_k is less than or equal to U', and
  - Assigning T_i to any processor in $\{P_1, P_2, \ldots P_{k-1}\}$ will make the total utilization of the processor to be greater than U'.

It can be proved that the first-fit algorithm achieves an approximation factor of 2, that is, the number of processors used by this algorithm is no more than twice the optimal number of processors. The proof proceeds with an important observation, that is, in a task assignment by the first-fit algorithm, there are no two processors that both have a utilization less than $U'/2$. This is because at any point of the process, if one processor is less than half-full (utilization $< U'/2$), we should assign any new task with a utilization less than $U'/2$ to that processor, instead of adding a new processor for the task. Therefore, if we have N_p processors, then at least $N_p - 1$ processors are more than half-full, i.e.,

$$\sum_{i=1}^{n} u_i > \frac{1}{2}(N_p - 1)U'$$

Let N^* be the optimal number of processors, then by the definition of U',

$$N^* \geq \frac{1}{U'} \sum_{i=1}^{n} u_i = \frac{1}{2}(N_p - 1)$$

Therefore, $N_p \leq 2N^*$.

Example 4.16 *First-Fit Algorithm*
Consider the tasks listed in Table 4.4. They are to be scheduled by EDF on multiple processors. The maximum utilization allocated for periodic tasks on each processor is 0.8. Table 4.5 shows how tasks are assigned to processors step by step, and the utilization of each processor is updated after each step of the process.

4.5.1.2 First-Fit Decreasing Algorithm
The first-fit decreasing algorithm is the same as the first-fit algorithm, except that in the first-fit decreasing algorithm, the tasks are first sorted in nonincreasing order according to their utilizations and then they are assigned in turn in that order. For example, the sorted list of the 10 tasks in Table 4.4 is $(T_{10}, T_5, T_2, T_8, T_1, T_3, T_4, T_6, T_7, T_9)$. Then, we can use the first-fit algorithm to assign them to processors based on this order.

4.5.1.3 Rate-Monotonic First-Fit (RMFF) Algorithm
Recall the theorem that RM can produce a feasible schedule for n_i periodic tasks if the total utilization of these tasks is not greater than

$$U_{RM}(n_i) = n_i(2^{1/n_i} - 1)$$

Table 4.4 A set of periodic and preemptable tasks

	T_1	T_2	T_3	T_4	T_5	T_6	T_7	T_8	T_9	T_{10}
p_i	10	8	15	20	10	18	9	8	20	16
e_i	2	2	3	4	3	2	1	2	2	5
u_i	0.2	0.25	0.2	0.2	0.3	0.11	0.11	0.25	0.1	0.31

Table 4.5 Task assignment by the first-fit algorithm.

Step	Task	Utilization	Assigned to	Postassignment utilization
1	T_1	0.20	P_1	$U_1 = 0.20$
2	T_2	0.25	P_1	$U_1 = 0.45$
3	T_3	0.20	P_1	$U_1 = 0.65$
4	T_4	0.20	P_2	$U_2 = 0.20$
5	T_5	0.30	P_2	$U_2 = 0.50$
6	T_6	0.11	P_1	$U_1 = 0.76$
7	T_7	0.11	P_2	$U_2 = 0.66$
8	T_8	0.25	P_3	$U_3 = 0.25$
9	T_9	0.10	P_2	$U_2 = 0.76$
10	T_{10}	0.31	P_3	$U_3 = 0.56$

Otherwise, the RM algorithm may not be able to produce a feasible schedule. The RMFF algorithm works by sorting tasks in a nondecreasing order according to their periods first. Then, we assign each task in turn in the sorted order, until all tasks are assigned in the first-fit manner. A task T_i is assigned to a processor if the total utilization of the x tasks that are already assigned to the processor and T_i is not greater than $U_{RM}(x + 1)$.

Example 4.17 *Task Assignment by RMFF*
Consider the tasks listed in Table 4.4 again. According to their periods, they are sorted as $(T_2, T_8, T_7, T_1, T_5, T_3, T_{10}, T_6, T_4, T_9)$. The step-by-step task assignment is listed in Table 4.6. The values of $U_{RM}(n)$ for $n = 2, 3, 4$, and 5 are also given in the table to help with the assignment decision.

4.5.2 Assignment with Communication Cost

The cost of communication between tasks is the time spent for them to communicate with each other. Denote by c_{ij} the time to communicate from T_i to T_j. If T_i and T_j are assigned to the same processor, c_{ij} is low and negligible. However, if they are assigned to different processors that are connected via some kind of network, c_{ij} may be significant. When we partition tasks to different processors that involve nontrivial communication cost, we should try to minimize the communication cost, as well as minimize the number of processors to use.

We assume a heterogeneous computing system in which the execution cost of a task depends on the processor that it is executed on. We further assume that the network is heterogeneous, that is, the cost of communication between two interacting tasks depends on the processors they are assigned to and the

Table 4.6 Task assignment by RMFF.

Step	Task	Utilization	Assigned to	Postassignment utilization
1	T_2	0.25	P_1	$U_1 = 0.25, n_1 = 1$
2	T_8	0.25	P_1	$U_1 = 0.50, n_1 = 2$
3	T_7	0.11	P_1	$U_1 = 0.66, n_1 = 3$
4	T_1	0.20	P_2	$U_2 = 0.20, n_2 = 1$
5	T_5	0.30	P_2	$U_2 = 0.50, n_2 = 2$
6	T_3	0.20	P_2	$U_2 = 0.7, n_2 = 3$
7	T_{10}	0.31	P_3	$U_3 = 0.31, n_3 = 1$
8	T_6	0.11	P_3	$U_3 = 0.42, n_3 = 2$
9	T_4	0.20	P_3	$U_3 = 0.62, n_3 = 3$
10	T_9	0.10	P_3	$U_3 = 0.72, n_3 = 4$

$U_{RM}(2) = 0.83, U_{RM}(3) = 0.78, U_{RM}(4) = 0.76, U_{RM}(5) = 0.74$

bandwidth of communication link. The objective is to minimize the sum of all the communication costs. The problem is also known to be NP-complete.

Let l_{ij} be the *interference cost* when T_i and T_j are placed on the same processor, incurred due to resource contention. If we do not want T_i and T_j to be assigned to the same processor, we can set l_{ij} to be a sufficiently large number. We allow different processors to have different maximum utilizations. Denote by U_k the maximum utilization of all tasks assigned to the k-th processor.

Assigning n periodic tasks into m processors with the objective to minimize the total communication cost can be formulated as an ILP problem:

$$\min \sum_{i=1}^{n} \sum_{j=1}^{n} \sum_{k=1}^{m} \sum_{l=1}^{m} (1 - d_{ij}) A_{ik} A_{jl} [c_{ij}(1 - d_{kl}) + l_{ij} d_{kl}] \tag{4.13a}$$

$$\text{subject to } A_{ik} \in \{0, 1\} \tag{4.13b}$$

$$\sum_{k=1}^{m} A_{ik} = 1, \quad \text{for } i = 1, 2, \ldots, n \tag{4.13c}$$

$$\sum_{i=1}^{n} A_{ik} u_i < U_k \quad k = 1, 2, \ldots, m \tag{4.13d}$$

$$d_{ik} = \begin{cases} 1, & if \ i = k \\ 0, & if \ i \neq k \end{cases} \tag{4.13e}$$

Equations (4.13b) and (4.13c) state that a task can only be assigned to one processor. $A_{ik} = 1$ if the task T_i is assigned to processor P_k; otherwise, $A_{ik} = 0$.

Equation (4.13d) states that the total utilization of all tasks assigned to a processor cannot exceed the maximum utilization of the processor. Equations (4.13a) and (4.13e) state that the total cost is the sum of the communication cost and interference cost between any two tasks T_i and T_j. The factor $(1 - d_{ij})$ excludes the case that $T_i = T_j$. $A_{ik}A_{jl}$ indicates that T_i is assigned to P_k and T_j is assigned to P_l. $c_{ij}(1 - d_{kl})$ counts the communication cost when $P_k \neq P_j$. $l_{ij}d_{kl}$ counts the interference cost when $P_k = P_j$.

Exercises

1 Each of the following systems contains independent and preemptable periodic tasks and is scheduled by the structured clock-driven scheduling algorithm. For each system, calculate its major cycle, choose an appropriate frame size, and construct the schedule of the first major cycle. If task slicing is necessary, it should be kept minimal.
   (a) $T_1 = (3, 1)$, $T_2 = (6, 1)$, $T_3 = (9, 2)$.
   (b) $T_1 = (4, 1)$, $T_2 = (6, 1)$, $T_3 = (8, 2)$, $T_4 = (12, 2)$.
   (c) $T_1 = (4, 1)$, $T_2 = (8, 1)$, $T_3 = (12, 2)$, $T_4 = (12, 3)$.
   (d) $T_1 = (3, 1)$, $T_2 = (6, 1)$, $T_3 = (8, 2)$, $T_4 = (12, 4)$.
   (e) $T_1 = (5, 1)$, $T_2 = (10, 2)$, $T_3 = (20, 5)$.

2 A system containing the following three independent and preemptable periodic tasks is scheduled according to the structured clock-driven scheduling algorithm:

$$T_1 = (3, 1), T_2 = (6, 2), T_3 = (12, 2).$$

   (a) Construct the schedule of the first major cycle.
   (b) An aperiodic task A with execution time 4.5 arrives at 1.5. Schedule A at the slack time intervals of the schedule produced in (a).
   (c) Reschedule A with slack stealing based on the schedule produced in (a).

3 Figure 4.10 shows a schedule of periodic tasks for two major cycles. The frame size of the scheduler is 4. Perform acceptance test for the following sporadic tasks:

Sporadic task	Release time	Execution time	Deadline
S_1	1.5	3	9
S_2	5	2	13
S_3	7	4	15
S_4	14	3.5	19

4 The following systems of independent and preemptable periodic tasks are scheduled by the RM algorithm. Construct the schedule in the segment of $(0, 25)$ for each system.
(a) $T_1 = (4, 1)$, $T_2 = (8, 2)$, $T_3 = (10, 3)$.
(b) $T_1 = (4, 1)$, $T_2 = (5, 1)$, $T_3 = (7, 1)$, $T_4 = (10, 1)$.
(c) $T_1 = (3, 1)$, $T_2 = (6, 1)$, $T_3 = (8, 2)$, $T_4 = (12, 4)$.

5 Consider the following systems of independent and preemptable periodic tasks that are scheduled by the RM algorithm. Test the schedulability of each system using the time demand analysis.
(a) $T_1 = (4, 1)$, $T_2 = (7, 2)$, $T_3 = (9, 2)$.
(b) $T_1 = (5, 1)$, $T_2 = (8, 2)$, $T_3 = (10, 2)$, $T_4 = (15, 2)$.
(c) $T_1 = (3, 1)$, $T_2 = (5, 1)$, $T_3 = (8, 2)$, $T_4 = (10, 1)$.
(d) $T_1 = (3, 0.5)$, $T_2 = (4, 1)$, $T_3 = (5, 2)$, $T_4 = (8, 1)$.

6 Redo the schedulability test in Problem 5 by plotting the time demand function of each task.

7 Schedule the following one-shot tasks with hard deadlines on a uniprocessor by the EDF algorithm. Assume that all tasks are independent and preemptable.

Task	Release time	Execution time	Deadline
T_1	0	2	6
T_2	1	3	9
T_3	3	2	15
T_4	4	1	8
T_5	5	4	12
T_6	8	2	11

8 Assume that the tasks in Problem 7 have precedence constraints shown in Figure 4.21. Construct the EDF schedule again. Can all tasks meet their deadlines?

9 The following systems of independent and preemptable periodic tasks are scheduled by the EDF algorithm. Construct the schedule in the segment of $(0, 15)$ for each system.
(a) $T_1 = (3, 1)$, $T_2 = (5, 3)$.
(b) $T_1 = (4, 1)$, $T_2 = (5, 1)$, $T_3 = (6, 3)$.
(c) $T_1 = (5, 1)$, $T_2 = (6, 3)$, $T_3 = (9, 2)$.

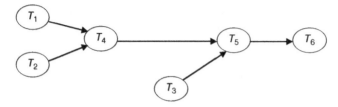

Figure 4.21 Precedence constraints graph of tasks in Problem 7.

10 Can we decide the schedulability of a system of independent and preemptable tasks by the RM algorithm based on its total utilization? Explain why.

11 Consider the following three independent periodic tasks:

$$T_1 = (4, 1), T_2 = (5, 2), T_3 = (10, 2).$$

T_1 and T_3 are preemptable, but T_2 is not.
(a) Construct the RM schedule of the system in the first 20 units of time.
(b) Construct the EDF schedule of the system in the first 20 units of time.

12 Consider four periodic tasks as follows:

$$T_1 = (3, 1), T_2 = (4, 1), T_3 = (6, 1), T_4 = (9, 1).$$

Assume the following:
- T_4 has a 0.2 time units of nonpreemptable portion, that is, $\theta_4 = 0.2$. All other tasks are preemptable at any time.
- T_2 may suspend itself at most once for 0.1 time units. That is, $k_2 = 1$ and $\rho_2 = 0.1$. All other tasks have no self-suspension.
- Overhead of context switches is not negligible; $CS = 0.1$.
Find out if the all tasks are schedulable by the RM algorithm.

13 Construct three periodic tasks that are schedulable by EDF and not all schedulable by RM. Assume that all tasks are independent and preemptable.

14 Construct three periodic tasks that are schedulable by RM if all of them are preemptable. However, if the task with the least priority is not preemptable, the task with the highest priority cannot meet its deadline. Assume that all tasks are independent.

15 Consider a system of two periodic tasks:

$$T_1 = (4.5, 2), T_2 = (6, 2).$$

They are scheduled by RM. An aperiodic task A with execution time 1.5 is released at 0.1. Construct the schedule of the system in the first 15 units of time in each of the following three cases:
(1) A is scheduled at the background of T_1 and T_2.
(2) A is scheduled using a simple polling server $T_s = (4, 0.5)$.
(3) A is scheduled using a deferrable server $T_{ds} = (4, 0.5)$.

16 Consider the assignment of the following periodic tasks to multiple processors:

	T_1	T_2	T_3	T_4	T_5	T_6	T_7	T_8	T_9	T_{10}
p_i	10	8	20	5	12	18	9	30	25	16
e_i	3	2	2	1	3	2	2	3	10	4
u_i	0.3	0.25	0.1	0.2	0.25	0.11	0.22	0.1	0.4	0.25

(a) Assume that the tasks are scheduled by EDF and the maximum utilization of each processor is 0.8. What is the assignment based on the first-fit algorithm?
(b) Assume that the tasks are scheduled by EDF and the maximum utilization of each processor is 0.75. What is the assignment based on the first-fit decreasing algorithm?
(c) What is the assignment based on the RM first-fit algorithm?

Suggestions for Reading

Liu and Layland introduced RM and showed that RM is optimal among all static scheduling algorithms in Ref. [1] in 1973. In a system where a task's relative deadline is less than its period, Leung and Whitehead showed that DM achieves the best performance in terms of schedulability [2]. Dertouzos showed that EDF is optimal among all online algorithms [3]. To schedule aperiodic tasks in fixed-priority systems, polling server and deferrable servers are introduced in Refs [4, 5], sporadic server in Ref. [6], and the slack stealer in Ref. [7]. For dynamic-priority systems, the dynamic sporadic server [8, 9], the total-bandwidth server [9], and the constant-bandwidth server [10] are the options.

References

1 Liu, C.L. and Layland, J.W. (1973) Scheduling algorithm for multiprogramming in a hard real-time environment. *Journal of the ACM*, **20** (1), 40–61.

2 Leung, J. and Whitehead, J. (1982) On the complexity of fixed priority scheduling of periodic real-time tasks. *Performance Evaluation,* **2** (4), 237–250.

3 Dertouzos, M.L. (1974) *Control Robotics: The Procedural Control of Physical Processes, Information Processing, vol. 74,* North-Holland Publishing Company, pp. 807–813.

4 Lehoczky, J.P., Sha, L. and Strosnider, J.K. 1987 Enhanced aperiodic responsiveness in hard real-time environments. Proceeding of the IEEE Real-Time Systems Symposium, pp. 261–270.

5 Strosnider, J.K., Lehoczky, J.P., and Sha, L. (1995) The deferrable server algorithm fro enhanced aperiodic responsiveness in hard real-time environments. *IEEE Transactions on Computers,* **44** (1), 73–91.

6 Sprunt, B., Sha, L., and Lehoczky, J.P. (June 1989) Aperiodic task scheduling for hard real-time systems. *Journal of Real-Time Systems,* **1,** 27–60.

7 Lehoczky, J.P. and Ramos-Thuel, R. 1992 An Optimal Algorithm for Scheduling Soft-Aperiordic Tasks in Fixed-Priority Preemptive Systems. Proceedings of the IEEE Real-Time Symposium

8 Ghazalie, T.M. and Baker, T.P. (1995) Aperiodic servers in a deadline scheduling environment. *Real-Time Systems,* **9** (1), 31–67.

9 Spuri, M. and Buttazzo, G. (1996) Scheduling aperiodic tasks in dynamic priority systems. *Real-Time Systems,* **10** (2), 179–210.

10 Abeni, L. and Buttazzo, G. 1998 Integrate multimedia applications in hard real-time systems, Proceedings of the IEEE Real-Time Systems Symposium, Madrid, Spain.

5

Resource Sharing and Access Control

The scheduling algorithms presented in Chapter 4 are all under the assumption that all tasks are independent. We remove this assumption in this chapter, because in many real-time applications, tasks do have explicit or implicit dependencies among them. Explicit dependencies can be specified by the task precedence graph, as discussed in Chapter 4. Data or resource sharing imposes implicit dependencies among tasks that share the resource. Many shared resources do not allow simultaneous access. When tasks share resources, scheduling anomalies may occur due to potential priority inversions and even deadlocks. This chapter discusses how resource sharing and resource contention affect the execution behavior and schedulability of tasks and how various resource access control protocols work to reduce the undesirable effect of resource sharing. We focus on priority-driven and single-processor systems.

5.1   Resource Sharing

Examples of common resources are data structures, variables, main memory area, files, registers, and I/O units. A task may need some resources in addition to a processor in order to make progress. For example, a computational task may share data with other computational tasks, and the shared data may be guarded by semaphores. Each semaphore is a resource. When a task wants to access the shared data guarded by a semaphore R, it must lock the semaphore first and then enters the critical section of the code where it accesses the shared data. In this case, we say that the task requires the resource R for the duration of the critical section.

We only consider *serially reusable* resources. A serially reusable resource is one that can be used safely by one task at a time and is not depleted by that use. If a task using the resource gets preempted, it is allowed to use it at some later time without a problem. Examples of serially reusable resources are devices, files, databases, and semaphores.

Real-Time Embedded Systems, First Edition. Jiacun Wang.
© 2017 John Wiley & Sons, Inc. Published 2017 by John Wiley & Sons, Inc.

Not all serially reusable resources are preemptable. A tape and CD are examples of nonpreemptable resources. If a process has begun to burn a CD-ROM, suddenly taking the CD recorder away from it and giving it to another process will result in a garbled CD. A task using a nonpreemptable resource cannot be preempted from the resource usage. In other words, when a unit of resource is granted to a task, this unit is no longer available to other tasks until the task frees the unit; otherwise, the resource may become inconsistent and lead to a system failure.

5.1.1 Resource Operation

Suppose that there are m types of resources in a system, namely $R_1, R_2, ..., R_m$, and there are v_i indistinguishable units of R_i, $i = 1, 2, ..., m$. When a task requests η_i units of a resource R_i, it executes a *lock* to request them. The action is denoted by $L(R_i, \eta_i)$. When the task no longer needs a resource, it executes an *unlock* to release the resource, denoted by $U(R_i, \eta_i)$. If the resource has only one unit, the notations are simplified to $L(R_i)$ and $U(R_i)$ for lock and unlock, respectively. A binary semaphore, for example, is a resource that has only one unit, while a counting semaphore can have multiple units.

The code segment of a task that begins at a lock and ends at a matching unlock is called a *critical section*, which cannot be concurrently executed by multiple processes. Resources are released in the last-in-first-out order. Therefore, critical sections are properly nested.

Two tasks *conflict* with each other if some of the resources they require are of the same type. They *contend* for a resource when one requests a resource that the other has. We say that the lock request $L(R_i, \eta_i)$ *fails* (or is *denied*) if the scheduler does not grant the η_i units of a resource R_i to the task. When its lock request is denied, the task is *blocked* and losses the processor. A blocked task is removed from the ready queue and stays blocked until the scheduler grants η_i units of R_i to it. At that time, the task becomes *unblocked* and is moved to the ready task queue.

5.1.2 Resource Requirement Specification

The duration of a critical section determines how long the task that locks the corresponding semaphore (resource) needs to hold the resource. We denote by $[R, \eta; c]$ that the task needs η units of a resource R and the execution time of the critical section is c. If only one unit of the resource is required, we omit the parameter η and use the simpler notation $[R; c]$ instead. Nested critical sections are denoted by nested brackets. For example, a notation by $[R_1; 10[R_2; 3]]$ means that the task requires one unit of R_1 for 10 units of time (because the critical section protected by R_1 is 10 units of time), while in the critical section of R_1, the task requires one unit of R_2 for 3 units of time.

Example 5.1 *Critical Sections*

Figure 5.1 shows the critical sections of two tasks, namely T_1 and T_2. The execution time of T_1 is 12 units of time. It has seven code segments:

[0, 2]: Requires no resources.
[2, 4]: Requires R_1.
[4, 6]: Requires no resources.
[6, 8]: Requires R_3.
[8, 10]: Requires R_3 and R_2.
[10, 11]: Requires R_3.
[11, 12]: Requires no resources.

The execution time of T_2 is 10 units of time. It has five code segments:

[0, 1]: Requires no resources.
[1, 4]: Requires R_2.
[4, 6]: Requires no resources.
[6, 8]: Requires R_1.
[8, 10]: Requires no resources.

The resource requirements of the two tasks are specified as follows:

T_1: $[R_1; 2], [R_3; 5[R_2; 2]]$.
T_2: $[R_2; 3], [R_1; 2]$.

For a periodic task T_i, by saying that T_i has a critical section $[R, \eta; c]$ we mean that every instance of the task has a critical section $[R, \eta; c]$.

5.1.3 Priority Inversion and Deadlocks

Assume that a higher priority task T_H and a lower priority task T_L share a resource R. T_H expects to run as soon as it is ready. However, if T_L is using the shared resource R when T_H becomes ready to run, T_H must wait for T_L to finish with it. We say that T_H is *pending* on the resource. This is a situation where priority inversion occurs: a lower priority task is running while a higher priority

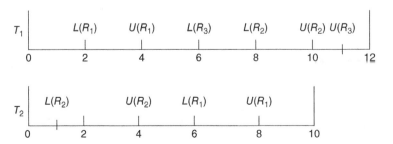

Figure 5.1 Critical sections.

task is waiting, which violates the priority model that a task can only be pre-empted by another task of higher priority. This is a *bounded* priority inversion. When T_L is finished using R and unlocks it, T_H preempts T_L and runs. Therefore, as long as the critical section of T_L with regard to R is not sufficiently long, T_H can still meet its deadline.

However, a much worse situation can occur. Let us say that after T_H is blocked by T_L on R and before T_L is done with R, a task T_{M1} with a priority between T_H and T_L is released and preempts T_L. Now T_L has to wait until T_{M1} is completed and then resumes its execution. While it is waiting, a task T_{M2} with a priority between T_H and T_{M1} is released and preempts T_{M1}; hence, T_{M1} has to wait until T_{M2} is completed and then resumes its execution. Such a chain waiting sequence can go on and on. We call this situation an *unbounded* priority inversion. An unbounded priority inversion can cause the task blocked on resource access to miss its deadline, as illustrated in Figure 5.2.

Tasks that share resources can also enter a state that none of them can make progress. Such a state is called a *deadlock*. A deadlock occurs when all the following four conditions are met:

- Mutual exclusion. One or more resources must be held by a task in an exclusive mode.
- Hold and wait. A task holds a resource while waiting for another resource.
- No preemption. Resources are not preemptable.
- Circular wait. There is a set of waiting tasks $T = \{T_1, T_2, ..., T_N\}$, such that T_1 is waiting for a resource held by T_2, T_2 is waiting for a resource held by T_3 and so on, until T_N is waiting for a resource held by T_1.

Figure 5.3 illustrates a deadlock of two tasks. The lower priority task T_L is released and executes first. It locks a resource A in a short period of time after it starts to run. Then, a higher priority task T_H is released and preempts T_L.

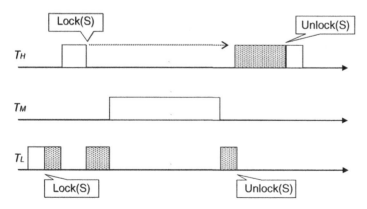

Figure 5.2 Priority inversion.

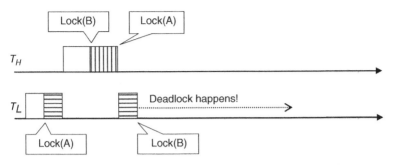

Figure 5.3 Deadlock.

Later on, T_H locks a resource B and continues running, until it attempts to lock the resource A and gets blocked, because A is held by T_L. Now, T_L executes. Sometime later, T_L attempts to lock the resource B, which is held by T_H. Hence, T_L is blocked. In other words, at this moment, both tasks are blocked and no one can move on.

5.1.4   Resource Access Control

Resource sharing can cause serious problems in real-time systems. We need rules to regulate the access to shared resources. Several well-known resource access control protocols have been developed to handle priority inversion and deadlocks caused by resource sharing. A resource access control protocol is a set of rules that govern when and under what conditions each request for resource is granted and how tasks requiring resources are scheduled. A well-designed access control protocol can prevent deadlocks from occurring. However, there is no protocol that can eliminate priority inversion. A realistic goal of access control is to keep the blocking time of higher priority tasks under control.

5.2   Nonpreemptive Critical Section Protocol

As we mentioned earlier, bounded priority inversion in general won't hurt an application provided that the critical section of the lower priority task executes in a timely manner. The real trouble is with unbounded priority inversion, in which a medium priority task preempts the lower priority task when the latter is executing its critical section. To prevent unbounded priority inversion from occurring, a simple way is to make all critical sections nonpreemptable. In other words, when a task locks a resource, it executes at a priority that is higher than the priorities of all other tasks, until it unlocks the resource (or completes the execution of its critical section). This protocol is called *nonpreemptive critical section* (NPCS) *protocol*. Because in this protocol, no task can be preempted when it holds a resource, circular waiting can never occur. Therefore, a deadlock is impossible.

Example 5.2   *Priority Inversion with Mars Pathfinder*
A well-known example of priority inversion is what happened with the Mars Pathfinder mission in July 1997. The Pathfinder mission was best known for the little rover that took high-resolution color pictures of the Martian surface and relayed them back to Earth. The problem was in the mission software run on the Martian surface. In the spacecraft, various devices communicated over a MIL-STD-1553 data bus. Activity on this bus was managed by a pair of high-priority tasks. One of the bus manager tasks communicated through a pipe with a low-priority meteorological science task (task *ASI/MET*).

The software mostly ran without incident on Earth. On Mars, however, a problem that was serious enough to trigger a series of software resets during the mission developed. The sequence of events leading to each reset began when the low-priority science task was preempted by a couple of medium-priority tasks while it held a mutex related to the pipe. While the low-priority task was preempted, the high-priority bus distribution manager (task *bc_dist*) tried to send more data to it over the same pipe. Because the mutex was still held by the science task, the bus distribution manager was made to wait. Shortly thereafter, the other bus scheduler became active. It noticed that the distribution manager hadn't completed its work for that bus cycle and forced a system reset.

Figure 5.4 illustrates the schedule of the Mars Pathfinder in which the higher priority task *bc_dist* misses its deadline due to unbounded priority inversion. However, if we reschedule these tasks with the NPCS protocol, *bc_dist* can

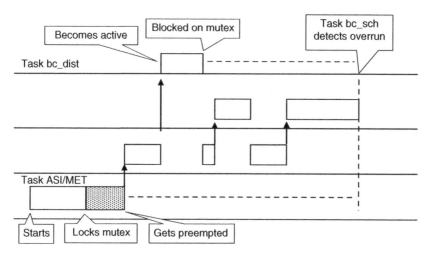

Figure 5.4 Priority inversion with Mars Pathfinder.

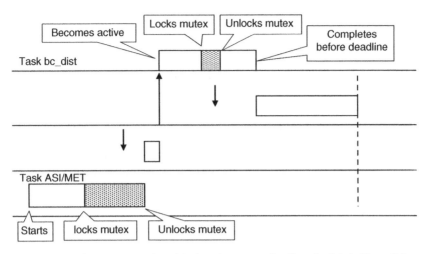

Figure 5.5 Apply nonpreemptive critical section protocol to the schedule in Figure 5.4.

complete before its deadline, as shown in Figure 5.5. In the new schedule, the task ASI/MET is able to complete its execution with the mutex without interruption. When the task *bc_dist* is ready to require the mutex, it is available. Therefore, there is no delay with the execution of *bc_dist*.

The NPCS protocol ensures that any higher priority task can only be blocked at most once. This is because, after being blocked by a lower priority task on a resource, the higher priority task starts execution immediately after the lower priority task is finished with the resource. After it starts running, it cannot be blocked by any lower level tasks. Therefore, in a system of periodic tasks T_1, T_2, ..., T_n that are indexed in order of nonincreasing priority, the maximum blocking time $b_i(rc)$ of T_i is

$$b_i(rc) = \max\{c_k, \quad k = i+1, i+2, \ldots, n\} \tag{5.1}$$

where c_k is the execution time of the longest critical section of the task T_k. The worst case (maximum blocking) occurs right after the lower priority task with the longest critical section enters the critical section, the higher priority task becomes ready and blocked.

Example 5.3 *Blocking Times of Tasks under the NPCS Protocol*
Consider a system of four periodic tasks T_1, T_2, T_3, and T_4. Their priorities are in the decreasing order. The execution times of their longest critical sections are 4, 3, 6, and 2, respectively. According to Eq. (5.1),

$$b_1(rc) = \max\{c_2, c_3, c_4\} = \max\{3, \ 6, \ 2\} = 6;$$

$$b_2(rc) = \max\{c_3, c_4\} = \max\{6,\ 2\} = 6;$$

$$b_3(rc) = \max\{c_4\} = \max\{2\} = 2.$$

Since T_4 has the lowest priority, no blocking on it can ever occur. Therefore, $b_4(rc) = 0$.

The NPCS protocol is simple and easy to implement. It works without needing any prior knowledge of resource requirements of tasks. It eliminates the possibility of unbounded priority inversion and deadlocks. However, under this protocol, any higher priority task can be blocked by any lower priority task that accesses some resources, even when the higher priority task does not need any resource in its entire execution.

5.3 Priority Inheritance Protocol

Another simple protocol eliminating unbounded priority inversion is the *Priority Inheritance Protocol*. Under this protocol, when a lower priority task blocks a higher priority task, it inherits the priority of the blocked higher priority task. This way the lower priority task can complete the execution of its critical section as soon as possible. Due to the priority inheritance, any task with a priority in between cannot preempt the lower priority task. Thus, unbounded priority inversion is avoided. A task holding a resource may block multiple tasks that are waiting for the resource. In that case, the last blocked task must have the highest priority and thus the blocking task inherits the highest priority. After executing its critical section and releasing the resource, the task returns to its original priority level.

Because the priority of a task is changeable under the priority inheritance protocol, we call the priority that is assigned to a task according to a scheduling algorithm (e.g., rate-monotonic) its *assigned priority*. In a fixed-priority scheduling system, the assigned priority of a task is a constant. Because a task can inherit other task's priority, a task may be scheduled at a priority that is different from its assigned priority. We can say that priority is the *current priority* of the task.

5.3.1 Rules of Priority Inheritance Protocol

There are three rules that govern the priority inheritance protocol:

Scheduling Rule: Ready tasks are scheduled on the processor preemptively according to their current priorities. A task's current priority is its assigned priority unless it is in its critical section and blocks higher priority tasks.

Table 5.1 Tasks in Example 5.4.

Task	Priority	Released at	Execution time	Resource usage
T_1	1	6	3	$[1, 2)$ uses X
T_2	2	4	5	$[1, 3)$ uses X; $[2, 3)$ uses Y
T_3	3	3	2	None
T_4	4	2	1	None
T_5	5	0	5	$[1, 4)$ uses X

Allocation Rule: When a task T locks a resource R at time t,
- If R is free (not locked by another task), R is allocated to T and the lock $L(R)$ is successful.
- If R is not free (locked by another task), $L(R)$ is denied and T is blocked.

Priority Inheritance Rule: When a lower priority task blocks a higher priority task, it inherits the *current* priority of the blocked higher priority task until it completes the execution of its critical section and unlocks the resource. Then, it returns to its assigned priority.

The reason that in the priority inheritance rule, we highlight the current priority to be inherited is because a task may block multiple tasks in the order of nondecreasing priorities, as is illustrated in the following example.

Example 5.4 *Priority Inheritance*

Consider the five single-shot tasks listed in Table 5.1. Their priority values are listed in the second column. As a convention, a greater priority value means a lower priority. Thus, T_1 has the highest priority and T_5 has the lowest priority. Their schedule is depicted in Figure 5.6. We explain the schedule step by step as follows.

At time 0, T_5 is released. It is the only task ready to execute. It starts execution.

At time 1, T_5 locks the resource X that is free. According to the allocation rule, T_5 is granted X. T_5 enters its critical section guarded by X.

At time 2, T_4 is released. Because the priority of T_4 is higher than that of T_5, T_4 preempts T_5.

At time 3, T_4 completes its execution. T_3 is released. Because the priority of T_3 is higher than that of T_5, T_3 is scheduled to run.

At time 4, T_2 is released. Because the priority of T_2 is higher than that of T_3, T_2 is scheduled to run. T_3 is preempted. Both T_5 and T_3 are waiting to run.

At time 5, T_2 attempts to lock X, which is held by T_5. According to the allocation rule, the lock is denied. Thus, T_2 is blocked by T_5. According to the priority inheritance rule, T_5 inherits T_2's priority as its current priority, which is higher than that of another waiting task T_3. Therefore, T_5 is scheduled to run. Both T_2 and T_3 are waiting.

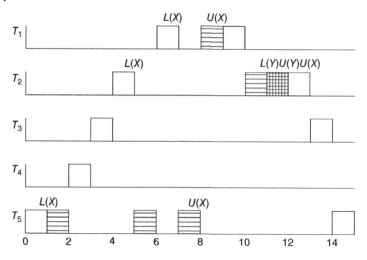

Figure 5.6 Schedule of tasks in Example 5.4.

At time 6, T_1 is released. Because T_1 has the highest priority in the system, it preempts T_5 and executes. T_2, T_3, and T_5 are waiting.

At time 7, T_1 attempts to lock X, which is held by T_5. According to the allocation rule, the lock is denied. Thus, T_1 is blocked by T_5. According to the priority inheritance rule, T_5 inherits T_1's priority as its current priority, which is higher than that of two waiting tasks T_2 and T_3. Therefore, T_5 is scheduled to run. T_1, T_2, and T_3 are waiting.

At time 8, T_5 completes the execution of its critical section and unlocks X. Its current priority drops to its assigned priority, the lowest in the system. Although it still has one unit to execute, it is preempted by T_1, which is granted X and starts to execute with X. T_2, T_3, and T_5 are waiting.

At time 9, T_1 completes the execution of its critical section and unlocks X. It still has one unit to execute. Because it has the highest priority in the system, so it continues running. T_2, T_3, and T_5 are waiting.

At time 10, T_1 completes its execution. Among the three waiting tasks T_2, T_3, and T_5, T_2 has the highest priority. Its pending lock of X is granted because X is free. Thus, T_2 executes. T_3 and T_5 are waiting.

At time 11, T_2 locks Y. Because Y is free, so the lock is successful. T_2 executes with both X and Y. T_3 and T_5 are waiting.

At time 12, T_2 unlocks X and Y. It continues running because it still has one time unit to finish. T_3 and T_5 are waiting.

At time 13, T_2 completes its execution. T_3 executes and T_5 is waiting.

At time 14, T_3 completes its execution. T_5 executes.

At time 15, T_5 completes its execution. All tasks are executed.

Throughout their executions, the current priorities of all tasks except T_5 are their assigned priorities, because they never block any other tasks. The current priority of T_5 changes as follows:

[0, 5): π_5; [5, 7): π_2; [7, 8): π_1; [8, 15): π_5.

Here we use notation π_i for the assigned priority of a task T_i. $\pi_i = i$ for $i = 1, 2, ..., 5$.

Discussion:

1. Notice that due to priority inheritance, T_3, a task whose priority is between T_2 and T_5 does not get chance to preempt T_5 when T_2 is blocked by T_5. Therefore, unbounded priority inversion is controlled.
2. T_2 and T_1 are *directly* blocked T_5 in turn at time 5 and 8, respectively, due to resource contention. We say that T_3 is *priority-inheritance* blocked by T_5 at time 5 and 7. Thus, there are two types of blocking in the priority inheritance protocol.

5.3.2 Properties of Priority Inheritance Protocol

We mentioned earlier that the NPCS protocol can avoid both unbounded priority inversion and deadlocks. We have already justified that the priority inheritance protocol can eliminate unbounded priority inversion. A natural question is: can the priority inheritance protocol eliminate deadlocks as well? Unfortunately, the answer is no. The reason is simple: it cannot avoid circular waiting of resources. Look at the deadlocked schedule depicted in Figure 5.3. This schedule does not violate any rules of the priority inheritance protocol. In other words, the priority inheritance protocol does not prevent deadlocks. We use another example to further illustrate this fact.

Example 5.5 *Deadlocks under Priority Inheritance Protocol*
Figure 5.7 shows the schedule of the three tasks listed in Table 5.2. At time 6, T_H locks the resource X that is in use by T_L, which is preempted in the critical section of X by T_M at time 2. According to the resource allocation rule, the lock fails and T_H is blocked. T_L's priority is upgraded to 1 and executes. At time 7, T_L locks the resource Y that is in use by T_H, so the locks fails. Now T_H and T_L enter a circular waiting for resources held by each other and neither of them can move on.

The other task, T_M, cannot run either, because its priority is not the highest. Therefore, a deadlock occurs.

A lower priority task blocks a higher priority task only when it is executing its critical sections. Otherwise, it is preempted by the higher priority task.

Under the priority inheritance protocol, a higher priority task can only be blocked for at most once by a lower priority task. This is because after the lower

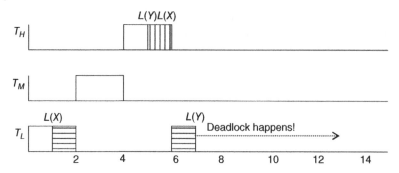

Figure 5.7 Schedule of tasks listed in Table 5.2.

Table 5.2 Tasks in Example 5.5.

Task	Priority	Released at	Execution time	Resource usage
T_H	1	4	4	$[1, 4)$ uses Y; $[2, 3)$ uses X
T_M	2	2	3	None
T_L	3	0	7	$[1, 5)$ uses X; $[3, 4)$ uses Y

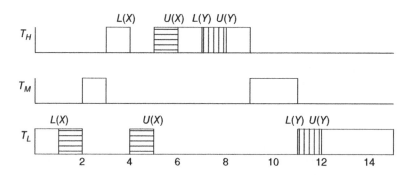

Figure 5.8 A higher priority task is blocked by a lower priority task for at most once.

priority task completes the execution of its critical section and unlocks the resource, the higher priority task is unblocked and executes. The lower priority task cannot run until the higher priority task completes its execution, even if the lower priority task has multiple critical sections. For example, in the schedule shown in Figure 5.6, T_1 is only blocked once by T_5, so does T_2. Figure 5.8 illustrates the case that the two tasks have conflict on two resources, but the higher priority task is only blocked once. Again, the reason is that, after T_L unlocks X at time 5, its current priority drops to its assigned priority, and it never gets a chance to run again before T_H and T_M complete.

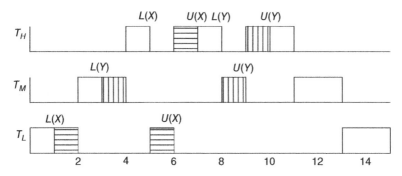

Figure 5.9 A higher priority task is blocked by multiple lower priority tasks.

A task can be blocked by multiple lower priority tasks. Consider the case illustrated in Figure 5.9. T_L and T_M lock X and Y at times 1 and 3, respectively. T_L is preempted by T_M at time 2, while T_M is preempted by T_H at time 4. T_H is blocked by T_L on X at time 5. Thus, T_L inherits T_H's priority and executes with X until it unlocks X at time 6, which unblocks T_H. T_H executes until it is blocked again by T_M on Y at time 8. It is unblocked when T_M unblocks Y at time 9.

The worst case of blocking under the priority inheritance protocol is that a task is blocked by every lower priority task for once by their longest critical sections.

In summary, the priority inheritance protocol eliminates unbounded priority inversion. A task can only be blocked by any lower priority task for at most once. However, there are several noticeable disadvantages with the protocol. First, it does not prevent deadlocks from occurring. Second, a task can be blocked by as many lower priority tasks as there are, so the blocking time can be long enough to cause the task to miss its deadline. Third, due to the existence of priority inheritance blocking, a task can be blocked by any lower priority task that has no resource conflict with it, a disadvantage that the NPCS protocol suffers as well.

Nevertheless, the priority inheritance protocol is supported by most commercial real-time operating systems. It was used to solve the priority inversion issue with the Mars Pathfinder back in 1997.

The priority inheritance protocol is also called the *basic* priority inheritance protocol. Several protocols were developed based on this protocol with better performance. The priority ceiling protocol to be presented in the next section is one of them.

5.4 Priority Ceiling Protocol

The priority ceiling protocol is an improvement of the priority inheritance protocol with the goal to prevent the formation of deadlocks and to reduce the blocking time. This protocol assumes that the resource requirement of every task is known before the execution of any task starts. It associates each resource with a *priority ceiling*, which is the highest priority of all tasks that might use the resource. As we know that in the priority inheritance protocol, whenever a request for a resource is made, the resource is granted to the requesting task as long as it is free. However, such a request under the priority ceiling protocol may or may not be granted, even if the resource is free. More specifically, when a task preempts the critical section of another task and locks a new resource, the protocol ensures that the lock is successful only if the priority ceiling of the new resource is higher than that of any preempted resources.

We use the following priority-related notations:

Π_R: the priority ceiling of a resource R

$\Pi(t)$: the *system's priority ceiling* at time t, which is the highest priority ceiling of all resources that are in use. If at time t, there are no resources in use, then the priority ceiling $\Pi(t)$ is the lowest, denoted by the symbol Ω.

5.4.1 Rules of Priority Ceiling Protocol

The rules for scheduling, resource allocation, and priority inheritance of the priority ceiling protocol are as follows:

Scheduling Rule: Ready tasks are scheduled on the processor preemptively according to their current priorities. A task's current priority is its assigned priority unless it is in its critical section and blocks higher priority tasks.

Allocation Rule: When a task T locks a resource R at time t,

- If R is not free (locked by another task), $L(R)$ is denied and T is blocked.
- If R is free,
  - If T's current priority $\pi(t)$ is higher than $\Pi(t)$, R is allocated to T and the lock $L(R)$ is successful.
  - If T's current priority $\pi(t)$ is lower than or equal to $\Pi(t)$, R is allocated to T only if T is the task holding the resource(s) whose priority ceiling is equal to $\Pi(t)$; otherwise, the lock fails and T is blocked.

Priority Inheritance Rule: When a lower priority task blocks a higher priority task, it inherits the *current* priority of the blocked higher priority task until it completes the execution of its critical section and unlocks the resource. Then, it returns to its assigned priority.

Notice that the scheduling rule and priority inheritance rule are exactly the same as they are in the priority inheritance protocol. The only difference is with the allocation rule. Before we explain the benefits of this new rule, let us use an example to show how the protocol works.

Table 5.3 Tasks in Example 5.6.

Task	Priority	Released at	Execution time	Resource usage
T_1	1	6	3	$[1, 2)$ uses X
T_2	2	4	5	$[1, 3)$ uses Y; $[2, 3)$ uses X
T_3	3	3	2	None
T_4	4	2	1	None
T_5	5	0	5	$[1, 4)$ uses X

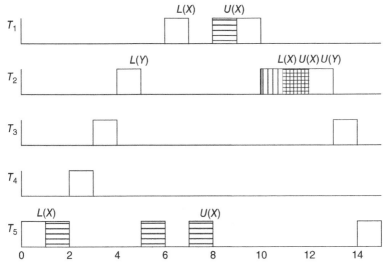

Figure 5.10 Schedule of tasks in Table 5.3.

Example 5.6 *Priority Ceiling Protocol*

The tasks listed in Table 5.3 are similar to those in Table 5.1. The only difference is with T_2, which uses X in $[1, 3)$ and Y in $[2, 3)$ in Table 5.1. The priority ceilings of the two resources are

$$\Pi_X = 1; \ \Pi_Y = 2.$$

The schedule of these tasks under the priority ceiling protocol is depicted in Figure 5.10.

When T_5 requests X at time 1, there is no other resource in use, so the priority ceiling $\Pi(1)$ is Ω, the lowest possible priority. According to the allocation rule, T_5 is granted X. Then T_5 is preempted by T_4 at time 2, T_4 is preempted by T_3 at time 3, and T_3 is preempted by T_2 at time 4.

At time 5, T_2 requests Y. At this moment, $\Pi(5) = \Pi_X = 1$. Because the priority of T_2 is less than $\Pi(5)$, the request is denied according to the resource allocation rule. Therefore, T_2 is blocked by T_5. T_5's current priority is upgraded to 2 and T_5 executes.

When T_1 is released at time 6, because its assigned priority is greater than T_5's current priority, it preempts T_5. At time 7, T_1 attempts to lock X and gets directly blocked by T_5. Thus, T_5's current priority is upgraded to 1 and T_5 executes.

T_5 unlocks X at time 8. T_1 is unblocked and executes its critical section of X. It unlocks X at time 9 and completes its execution at time 10. Notice that T_2 is also unblocked at time 8. However, because its current priority is lower than that of T_1, it is not picked by the scheduler to run.

After T_1 completes its execution at time 10, T_2 executes. At time 11, it requests X. At this moment, $\Pi(11) = \Pi_Y = 2$, which is equal to the current priority of T_2. Because T_2 is the task that holds Y, therefore, according the allocation rule, T_2 is granted X. T_2 completes at time 13. Then, T_2 and T_5 complete the executions of their remaining portions at time 14 and 15, respectively.

Throughout their execution, the current priorities of all tasks except T_5 are their assigned priorities, because they never block any other tasks. The current priority of T_5 changes as follows:

$$[0, 5): \pi_5; [5, 7): \pi_2; [7, 8): \pi_1; [8, 15): \pi_5.$$

The system's priority ceiling changes as follows:
$$[0, 1): \Omega; [1, 9): \pi_1; [10, 11): \pi_2; [11, 12): \pi_1; [12, 15): \Omega.$$

Discussion:

1. When T_2 requests Y at time 5, Y is free. Under the priority inheritance protocol, such a request would be granted. So, the priority inheritance protocol is a greedy protocol. Because it is greedy, it does not prevent deadlocks.
2. In addition to the direct blocking and priority inheritance blocking, there is a third type of blocking under the priority ceiling protocol, which is *priority ceiling blocking*. In Example 5.6, the priority ceiling blocking occurs at time 5 when T_2 requests Y.

5.4.2 Properties of Priority Ceiling Protocol

The priority inheritance rule allows the priority ceiling protocol to avoid unbounded priority inversion, a property shared by the priority inheritance protocol. As we see in Example 5.6, the tough resource allocation rule in the priority ceiling protocol causes additional priority ceiling blocking. This additional blocking, however, helps avoid deadlocks. Therefore, priority ceiling blocking is also called deadlock avoidance blocking. Example 5.7 shows how it works.

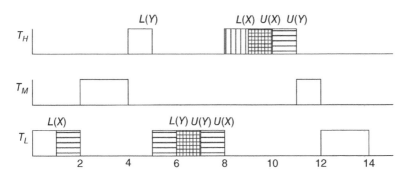

Figure 5.11 The priority ceiling protocol avoids deadlock.

Example 5.7 *Deadlock Avoidance Blocking*

We know that the three tasks listed in Table 5.2 enter a deadlock when they are scheduled with the priority inheritance protocol. Let us see what happens if we reschedule them with the priority ceiling protocol. The priority ceilings of the two resources are

$$\Pi_X = 1; \Pi_Y = 1.$$

As shown in Figure 5.11, the first resource request is made by T_L at time 1 when the system's priority ceiling is Ω. The request is granted. Then, T_L is preempted by T_M at time 2, and T_M is preempted by T_H at time 4. At time 5, T_H requests Y. Because X is in use, the system's priority ceiling $\Pi(5)$ is 1, which is equal to the current priority of T_H. Recall the allocation rule that states that in this case, if the resource whose priority ceiling is equal to the requesting task's current priority is held by the requesting task, the request is granted; otherwise, it is denied. Here, the resource X is held by T_L, not T_H. Therefore, the request is denied, and T_H is priority-ceiling blocked. T_L inherits T_H's priority and runs until it unlocks X at time 8. Then, T_H resumes its execution until it is completed at time 11. T_M and T_L complete their remaining portions at 12 and 14, respectively. The deadlock is avoided!

The priority ceiling protocol prevents the formation of deadlocks by denying those resource requests that, if granted, may cause deadlock later. If the current priority of the requesting task is higher than the system's priority ceiling, then based on the definition of the system's priority ceiling, we know for sure that the requesting task will not request any resource that is in use. So, it is safe to allocate the resource to the task. Otherwise, the requesting task may need to use some resource that is in use by other tasks later, which can lead to a deadlock, so the request should be denied. An exception is that requesting task itself holds the resource whose priority ceiling is equal to the system's priority ceiling. In that case, the requesting task will not request any resource that is in use by other tasks. So, it is safe to allocate the resource to the task.

In addition to deadlock avoidance, the priority ceiling protocol can prevent the chained blocking illustrated in Figure 5.9. The tasks shown in Figure 5.9 are rescheduled with the priority ceiling protocol, and the new schedule is depicted in Figure 5.12. In the new schedule, T_M is priority-ceiling blocked at time 3 when it locks Y. As a result, the blocking on T_H by T_M due to resource contention for Y is avoided. (In this particular example, the blocking by T_L is also avoided, but this is not the general case. If the length of the critical section of T_L is longer than 2 time units, then T_H will be blocked by T_L at time 5.)

In fact, as long as one resource that a higher priority task will access is in use by a lower priority task, no tasks whose priorities are in between can successfully lock any resource that the higher priority task will access. In other words, under the priority ceiling protocol, a task can be blocked for at most the duration of one critical section.

In summary, the priority ceiling protocol eliminates unbounded priority inversion, prevents the formation of deadlock, and avoids chained blocking. A task can only be blocked for at most the duration of one critical section. These advantages show that it is a big improvement over the priority inheritance protocol. Of course, the improvement comes at a cost, which includes the overhead of the calculation of resource priority ceiling and system priority ceiling, and the overhead of extra context switches due to priority ceiling blockings. In general, a task that does not need to access resources may suffer at most two context switches due to preemption. A task that accesses resource may suffer additional two context switches incurred by blockings.

5.4.3 Worst-Case Blocking Time

As we mentioned earlier, there are three types of blocking when resource accesses are controlled by the priority inheritance protocol: direct blocking, priority inheritance blocking, and priority ceiling blocking.

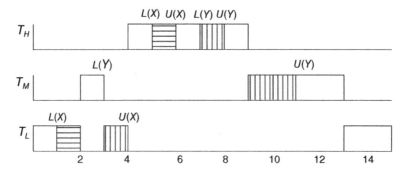

Figure 5.12 The priority ceiling protocol avoids chained blocking.

Direct blocking occurs when a higher priority task requests a resource that is held by a lower priority task. The maximum blocking time is the duration of the corresponding critical section of the lower priority task.

When a lower priority task blocks a higher priority task, it inherits the priority of the higher priority task. Thus, it further priority-inheritance blocks all tasks whose priorities are between the assigned priorities of the two tasks. The maximum blocking time is the duration of the corresponding critical section of the lower priority task.

After a lower priority task successfully locks a resource R, every higher priority task is priority-ceiling blocked from locking other resources if its priority is not higher than the priority ceiling of R. The maximum blocking time is the duration of the corresponding critical section of the lower priority task.

Notice that direct blocking and priority ceiling blocking never occur to tasks that do not use resources, as these two types of blockings only occur when tasks request resources. Besides, even if it conflicts with other tasks on multiple resources, a task can only be blocked for at most the duration of one critical section. Next, we use an example to show how to compute the worst-case blocking time of a task under the priority ceiling protocol.

Example 5.8 *Blocking Time Computation*
Consider a system of five tasks and three resources. The resource requirements of the tasks are specified as follows:

$T_1 : [X; 2][Y; 4]$

$T_2 : [Z; 1]$

$T_3 : [Y; 3][Z; 6]$

$T_4 : $ None

$T_5 : [X; 4][Z; 2]$

These tasks are indexed in the decreasing order of priorities. Let us discuss all possible blockings and blocking times.

Direct blockings:
- The lowest priority task T_5 shares X with T_1 and Z with T_2 and T_3, and thus, T_5 can directly block T_1 for 4 time units (the duration of T_5's critical section of X), and T_2 and T_3 each for 2 time units (the duration of T_5's critical section of Z).
- T_3 shares Y with T_1 and Z with T_2, and thus, T_3 can directly block T_1 for 3 time units (the duration of T_3's critical section of Y) and T_2 for 6 time units (the duration of T_3's critical section of Z).

Table 5.4 Worst-case blocking times of tasks in Example 5.8.

	Direct				Priority-inheritance				Priority-ceiling			
Task	T_2	T_3	T_4	T_5	T_2	T_3	T_4	T_5	T_2	T_3	T_4	T_5
T_1		3		4								
T_2		6		2		3		4		3		4
T_3				2				4				4
T_4								4				

Priority inheritance blockings:

- After T_5 blocks T_1, it can priority-inheritance block T_2, T_3 and T_4 for 4 time units in the worst case.
- After T_5 blocks T_2, it can priority-inheritance block T_3 and T_4 for 2 time units in the worst case.
- After T_5 blocks T_3, it can priority-inheritance block T_4 for 2 time units in the worst case.
- After T_3 blocks T_1, it can priority-inheritance block T_2 for 3 time units in the worst case.

Priority ceiling blockings:

- After T_5 locks X successfully, it can priority-ceiling block T_2 and T_3 for 4 time units in the worst case.
- After T_5 locks Z successfully, it can priority-ceiling block T_3 for 2 time units in the worst case.
- After T_3 locks Y successfully, it can priority-ceiling block T_2 for 3 time units in the worst case.

Table 5.4 shows the all different types of blockings and maximum blocking times. The table is composed of three subtables, and they are for directly blocking, priority inheritance blocking and priority ceiling blocking. There is a row for each task that can be blocked. There is no row for T_5 because it has the lowest priority and thus never gets blocked. For example, the row for T_1 shows that T_1 can be directly blocked by T_3 and T_5 for 3 and 4 time units, respectively. It also shows that there is neither priority inheritance nor priority ceiling blocking on T_1. The row for T_3 shows that T_3 can be directly blocked by T_5 for 2 time units, priority-inheritance blocked by T_5 for 4 time units, and priority-ceiling blocked by T_5 for 4 time units. We can also read the table in columns. For example, the column for T_5 in the priority inheritance blocking subtable shows that T_5 can priority-inheritance block each of T_2, T_3, and T_4 for 4 time units.

The values of most entries in the table come directly for the analysis that we just made. However, in case that there are multiple values for an entry, we should put the greatest one there because we are considering the longest blocking times. For example, our aforementioned analysis indicates that T_3 can be priority-inheritance blocked by T_5 for 4 time units after T_5 directly locks T_1. It can also be priority-inheritance blocked by T_5 for 2 time units after T_5 directly blocks T_2. We put 4 on the corresponding entry in the priority inheritance blocking subtable.

Because a task can only be blocked for at most the duration of one critical section, the maximum blocking time of a task is equal to the maximum value of entries in the corresponding row, which gives

$$b_1(rc) = 4; \quad b_2(rc) = 6; \quad b_3(rc) = 4; \quad b_4(rc) = 4.$$

In general, the direct blocking table is constructed based on the resource requirement specification. The entry (i, j) of the priority inheritance table is the maximum of all entries in the column j and rows $1, 2, ..., i-1$ of the direct blocking table. When the priorities of all tasks are distinct, the entries in priority ceiling blocking table are the same as the entries in the priority inheritance blocking table, except for tasks that do not require any resources, which are never priority-ceiling blocked.

5.5 Stack-Sharing Priority Ceiling Protocol

The stack-sharing priority ceiling protocol is a protocol that provides stack-sharing capability and simplifies the priority ceiling protocol. The worst-case blocking time of a task under this protocol is the same as it is under the priority ceiling protocol, but this protocol incurs smaller overhead of context switches compared to the priority ceiling protocol.

Sharing of the stack among tasks eliminates stack space fragmentation and thus saves memory. Normally, each task has its own runtime stack that stores its local variables and return address, as illustrated in Figure 5.13a. The memory used by a task stack is automatically allocated when the task is created. When the number of tasks in a system is too large, it may be necessary for several tasks to share a common runtime stack to reduce the overall memory demand, as illustrated in Figure 5.13b.

5.5.1 Rules of Stack-Sharing Priority Ceiling Protocol

When tasks share a common runtime stack, the task that executes is the one on the top of the stack. The space for the task is freed when it completes. When a task T_i is preempted by another task T_j, T_j is allocated the space above that of T_i. A preempted task can only resume when it returns to be on the top of the stack, that is, all tasks holding stack space above its space complete. Obviously, such a principle of space sharing does not allow any kind of blockings to occur.

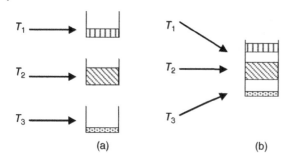

Figure 5.13 Runtime stacks of tasks. (a) No stack sharing. (b) Stack sharing.

(a) (b)

Imagine that after it preempts T_i, if T_j requests a resource that is held by T_i, then T_j on the top of the stack is blocked while the blocking task T_i cannot run before it is not on the top. Then, a deadlock occurs. Therefore, when a task is scheduled to run, we need to make sure that it does not get blocked due to resource access before it completes.

The rules that define the stack-sharing priority ceiling protocol are as follows:

Update of Priority Ceiling: When all the resources are free, the priority ceiling of the system $\Pi(t)$ is Ω. $\Pi(t)$ is updated each time a resource is allocated or freed.

Scheduling Rule: After a task is released, it is blocked from starting execution until its assigned priority is higher than $\Pi(t)$. At all times, tasks that are not blocked are scheduled on the processor in a priority-driven, preemptive fashion according to their assigned priorities.

Allocation Rule: Whenever a task requests a resource, it is allocated the resource.

Example 5.9 *Stack-Sharing Priority Ceiling Protocol*

We reschedule the five tasks listed in Table 5.3 with the stack-sharing priority ceiling protocol. The new schedule is depicted in Figure 5.14. It is much simpler than the priority ceiling protocol-based schedule shown in Figure 5.10. It is explained as follows.

At time 0, T_5 is released and the stack is empty, so it executes.

At time 1, T_5 requests X. According to the allocation rule, T_5 is allocated X. The system's priority ceiling $\Pi(t)$ is upgraded from Ω to Π_X, which is equal to 1.

At time 2, T_4 is released. Because the assigned priority of T_4 is lower than $\Pi(t)$, it is blocked. T_5 continues its execution.

At time 3, T_3 is released. Because the assigned priority of T_3 is lower than $\Pi(t)$, it is blocked. T_5 continues to run.

At time 4, T_5 unlocks X. $\Pi(t)$ becomes Ω. Meanwhile, T_2 is released. Because the assigned priority of T_2 is higher than $\Pi(t)$, it is scheduled to run. T_5 is preempted. A stack space is allocated for T_2 on top of T_5.

At time 5, T_2 requests Y, which is free. According to the allocation rule, T_2 is allocated Y. $\Pi(t)$ is upgraded from Ω to Π_Y, which is equal to 2.

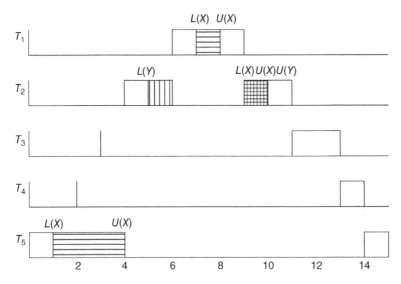

Figure 5.14 Stack-sharing schedule of tasks in Table 5.3.

At time 6, T_1 is released. Because the assigned priority of T_1 is higher than $\Pi(t)$, it is scheduled to run. T_2 is preempted. A stack space is allocated for T_1 on top of T_2. Now T_5 sits in the bottom of the shared stack, T_2 is in the middle, and T_1 is running on the top.

At time 7, T_1 requests X, which is free. T_1 is allocated X. $\Pi(t)$ is upgraded from Π_Y to Π_X, which is equal to 1. T_1 continues to run.

At time 8, T_1 unlocks X. $\Pi(t)$ drops to Π_Y. T_1 continues to run.

At time 9, T_1 completes. T_2 returns to be on the top of the stack. At this moment, underneath T_2 is T_5. There are also two blocked tasks: T_3 and T_4. Since T_2 has the highest priority, so T_2 executes. T_2 requests and is allocated X. $\Pi(t)$ is upgraded from Π_Y to Π_X.

At time 10, T_2 unlocks both X and Y. $\Pi(t)$ drops to Ω. T_2 continues to run.

At time 11, T_2 completes. T_5 returns to be on the top of the stack. The two blocked tasks T_3 and T_4 are also waiting to run. Since T_3 has the highest priority, T_3 is allocated a stack space on top of T_5 and executes.

At time 13, T_3 completes. T_5 returns to be on the top of the stack. However, because T_4 has a higher priority than T_5, T_4 is allocated a stack space on top of T_5 and executes.

At time 14, T_4 completes. The only reaming task T_5 executes and completes at time 15.

5.5.2 Properties of Stack-Sharing Priority Ceiling Protocol

The stack-sharing priority ceiling protocol is an improved version of the basic priority ceiling protocol that was introduced in the previous section.

It possesses all advantages that the priority ceiling protocol offers: no unbounded priority inversion, no deadlocks, and no chained blocking. Blockings under the stack-sharing priority ceiling protocol can only occur when tasks are released. Once a task starts execution, it can only be preempted but never get blocked before it completes. Because of this, no task ever suffers more than two context switches. This is compared to the priority ceiling protocol where a task that needs to access resources may suffer four context switches: two caused by preemption and two caused by blocking.

Although the stack-sharing priority ceiling protocol was developed with stack sharing in consideration, it can be reformulated for non-stack-sharing systems. The rules of the reformulated protocol are defined as follows:

Scheduling Rules:

1. Every task executes at its assigned priority when it does not hold resources.
2. Tasks of the same priority are scheduled on FIFO basis.
3. The priority of each task holding resources is the highest of the priority ceilings of all resources held by the task.

Allocation Rule: Whenever a task requests a resource, it is allocated the resource.

Exercises

1 What is priority inversion? In a priority-driven system where all tasks are preemptable and that only need the processor to run, will priority inversion occur? Among the four protocols introduced in this chapter, which protocols prevent priority inversion?

2 What is a deadlock? What are the conditions under which a deadlock can occur? Does the NPCS protocol prevent deadlocks? Does the priority inheritance protocol prevent deadlocks?

3 Explain each of the following terminologies:
   (a) Assigned priority
   (b) Current priority
   (c) Resource priority ceiling
   (d) System priority ceiling
   (e) Direct blocking
   (f) Priority inheritance blocking
   (g) Priority ceiling blocking

4 How would you evaluate the performance of a resource access control protocol?

5 Determine whether the following statements are true or false. Justify your answer briefly.

   (a) A higher priority task can block a lower priority task in the NPCS protocol when they share a common resource.

   (b) In the priority inheritance protocol, a task that holds a resource has the highest priority of all tasks in the system.

   (c) Priority ceiling blocking is also called avoidance blocking, because the blocking prevents deadlock among tasks.

   (d) The fundamental difference between the priority inheritance protocol and priority ceiling protocol is that they use different priority inheritance rules.

   (e) Under the stack-sharing priority ceiling protocol, a task that does not require any resource does not undergo priority inversion.

   (f) Under the priority ceiling protocol, the maximum blocking time of a task is the sum of its direct blocking time, priority inheritance blocking time, and priority ceiling blocking time.

6 The following five tasks are to be scheduled on a single processor preemptively.

Task	Priority	Released at	Execution time	Resource usage
T_1	1	7	3	$[1, 2)$ uses X
T_2	2	5	3	$[1, 3)$ uses X; $[2, 3)$ uses Y
T_3	3	4	2	None
T_4	4	2	2	$[1, 2)$ uses X
T_5	5	0	5	$[1, 4)$ uses X

   (a) Construct the schedule when there is no resource access control.

   (b) Construct the schedule when the NPCS protocol is applied to control the access to resources. Explain the schedule step by step.

   (c) Construct the schedule when the priority inheritance protocol is applied to control the access to resources. Explain the schedule step by step.

7 The following five tasks are to be scheduled on a single processor preemptively.

Task	Priority	Released at	Execution time	Resource usage
T_1	1	8	2	[1, 2) uses X
T_2	2	5	4	[2, 3) uses Y
T_3	3	3	2	[1, 2) uses X
T_4	4	2	1	None
T_5	5	0	5	[1, 4) uses X

(a) Construct the schedule when the priority inheritance protocol is applied for resource access control. Explain the schedule step by step.

(b) Construct the schedule when the priority ceiling protocol is applied for resource access control. Explain the schedule step by step.

8 The following four tasks are to be scheduled on a single processor preemptively.

Task	Priority	Released at	Execution time	Resource usage
T_1	1	8	2	[1, 2) uses X
T_2	2	5	4	[2, 3) uses Y
T_3	3	3	2	[1, 2) uses X
T_4	4	2	1	None
T_5	5	0	5	[1, 4) uses X

(a) Construct the schedule when the priority ceiling protocol is applied to control access to resources. Explain the schedule step by step.

(b) Construct the schedule when the stack-sharing priority ceiling protocol is applied to control access to resources. Explain the schedule step by step.

9 A system contains five tasks. There are three resources X, Y, and Z. The resource requirements of the tasks are specified as follows:

T_1: [X; 1] [Y; 2].
T_2: [Z; 4].
T_3: [X; 2].
T_4: [X, 2][Y; 3].
T_5: [X; 1] [Z; 3].

The tasks are indexed in the decreasing order of priorities and scheduled with the priority ceiling protocol. Find the maximum blocking time for each task.

10 A system contains seven tasks. There are three resources X, Y, and Z. The resource requirements of the tasks are specified as follows:

T_1: $[Y; 2]$.
T_2: $[Z; 1]$.
T_3: $[X; 2]$.
T_4: $[X, 2]$ $[Y; 3]$.
T_5: $[Y; 1]$ $[Z; 3]$.
T_6: $[X; 2]$.
T_7: $[X, 2]$ $[Z; 2]$.

The tasks are indexed in the decreasing order of priorities and scheduled with the priority ceiling protocol. Find the maximum blocking time for each task.

Suggestion for Reading

Lampson and Redell published one of the first papers to point out the priority inversion problem [1]. The priority inheritance protocol and priority ceiling protocol were introduced by Sha, Rajkumar, and Lehoczky [2]. Baker proposed the stack-based resource allocation policy for real-time processes in Ref. [3]. Chen and Lin implemented the priority ceiling protocol in dynamic priority systems [4]. Chen and Ras described the implementation of the priority ceiling protocol in Ada-2005 [5]. A comprehensive review of priority inversion control policies is provided in Ref. [6].

The four Coffman conditions for deadlock to occur are described in Ref. [7] as well as many other operating system textbooks. Davison and Lee discussed deadlock prevention in concurrent real-time systems in Ref. [8].

References

1 Lampson, B. and Redell, D. (1980) Experience with processes and monitors in MESA. *Communications of the ACM*, **23** (2), 105–117.
2 Sha, L., Rajkumar, R., and Lehoczky, J. (1990) Priority inheritance protocols: an approach to real-time synchronization. *IEEE Transactions on Computers*, **39** (9), 1175–1185.
3 Baker, T.P. (1991) A stack-Based Resource Allocation Policy for Real-Time Processes. *Proceedings of the 12th IEEE Real-Time Systems Symposium*, San Antonio.
4 Chen, M.L. and Lin, K.J. (1990) Dynamic priority ceilings: A concurrency control protocol for real-time systems. *Real-Time Systems Journal*, **2** (4), 325–346.

5 Cheng, A. and Ras, J. (2007) The implementation of the Priority Ceiling Protocol in Ada-2005. *ACM SIGADA Ada Letters*, **27** (1), 24–39.

6 Liu, J. (2000) *Real-Time Systems*, Prentice Hall.

7 Silberschatz, A. (2006) *Operating System Principles*, 7th edn, Wiley-India.

8 Davidson, S. and Lee, I. (1993) Deadlock prevention in concurrent real-time systems. *Real-Time Systems*, **5**, 305–318.

6

Concurrent Programming

The previous chapters covered the concepts and theory of real-time tasks, task scheduling, and resource access control. This chapter shifts to the implementation side of real-time embedded software, with the focus being placed on the mechanisms of intertask synchronization and communication. All programming examples given in this chapter are tested in the Code::Blocks integrated development environment on Ubuntu.

6.1 Introduction

Concurrent programming is a technique for expressing potential parallelism. It divides an overall computation into subcomputations that may be executed concurrently, that is, several computations are executing during overlapping time periods.

Virtually, all real-time systems are inherently concurrent – devices operate in parallel in the real world. For example, a washing machine has to do several things, frequently more than one at a time: accepting water, monitoring the water level, monitoring the water temperature, timing a wash cycle, releasing the detergent, controlling the agitator, spinning the tub, and so on. A thermostat has at least three tasks running all the time: monitoring the room temperature, monitoring the timer, and polling the keypad. Therefore, a natural way to programming the controllers of these devices is use of concurrency.

Concurrent programming offers several distinguished advantages to programmers and application users. Firstly, it improves the responsiveness of an application. With concurrent computation, the system appears to immediately respond to every user request, even if the system is executing other expensive computation. Secondly, it improves the processor utilization. Multiple tasks compete for the processor time and keep the processor busy whenever a task

Real-Time Embedded Systems, First Edition. Jiacun Wang.
© 2017 John Wiley & Sons, Inc. Published 2017 by John Wiley & Sons, Inc.

is ready to run. If one task gets stuck, others can run. Thirdly, it provides a convenient structure for failure isolation.

You might wonder whether we can use sequential programming techniques to handle concurrent tasks. Well, if you choose to do that, you must construct the system so that it involves the cyclic execution of a program sequence to handle the various concurrent activities that are essentially irrelevant to each other. The resulting programs will be more obscure and inelegant and thus make the decomposition of the problem more complex.

It is worth mentioning that concurrent programming and parallel programming are two related concepts, but they are different. The goal of concurrent programming is to allow tasks to multiplex their executions on a single processor and tackle the complexity of problem-solving. However, at any given time, only one task is running in concurrent computing.

By contrast, execution of multiple tasks in parallel computing literally occurs at the same instant, for example, on separate processors of a multiprocessor machine, with the goal of speeding up computations. Parallel computing is impossible on a single processor, as only one computation can occur at any instant.

6.2 POSIX Threads

In this chapter, POSIX threads, or *Pthreads*, are chosen as an example to show the issues and their solutions in concurrent programming. The reason is that POSIX provides a set of standard application programming interfaces (APIs) for real-time embedded systems. Although many RTOS vendors have their proprietary programming interfaces, POSIX conformance is required, which permits interoperability of software across different operating systems and hardware implementations. Notice that the goal of this chapter is not teaching Pthreads, so we are not digging into the details on Pthreads. Instead, only routines that are necessary to explain concurrent programming concepts are briefly introduced. Readers who have programming experience in C, C++, or Java should be able to understand the programs listed in this chapter.

Pthreads are defined as a set of C language programming types and procedure calls, implemented with a `pthread.h` header file and a thread library. Pthreads API routines are provided for thread management, mutexes, condition variables, and thread synchronization with read/write locks and barriers.

Table 6.1 Pthread attributes.

Attribute	Description
Scope	The scope that the thread competes for resource
Detachstate	Thread is joinable or not
Stackaddr	Stack space allocated for the thread
Stacksize	Stack size of the thread
Priority	Priority of the thread
Policy	Scheduling policy for the thread
Inheritsched	Whether scheduling policy and parameters are inherited from parent thread
Guardsize	Size of the guard area for the stack of the thread

A new thread is created by calling the `pthread_create()` routine, which has the following prototype:

```
int pthread_create(pthread_t *thread,
        const pthread_attr_t *attr,
        void * (*start_routine) (void *),
        void *arg);
```

The first argument `thread` of the routine is an opaque and unique identifier of the new thread. The `attr` argument points to a `pthread_attr_t` structure whose contents are used to determine the attributes for the new thread. The new thread starts execution by invoking `start_routine()`; `arg` is passed as the sole argument of `start_routine()`. Upon success, the call returns 0; upon receiving an error, it returns an error number.

Attributes are used to specify the behavior of threads. Table 6.1 lists all attributes and their brief descriptions.

The default attribute object is NULL, in which the stack size is typically set to 1 MB, the thread priority is set to the priority of its parent thread, and the scheduling policy is time-sharing, for example. Attributes are specified only at thread creation time; they cannot be altered after the thread is created and running. An attribute object is opaque and cannot be directly modified by assignments. A set of functions is provided to initialize, configure, and destroy each attribute object. For example, the `pthread_attr_setscope(attr)` routine is used to set the scope of the specified attribute. The type of attribute object is pthread_attr_t. The routine that initializes an attribute object is pthread_attr_init(attr). When the created thread is no longer needed, the memory of the attribute object should be freed up by calling the pthread_attr_destroy(attr) routine.

Other important thread management routines include the following:

- void pthread_exit(void *status) This routine is called to terminate the calling thread.
- int pthread_join(pthread_t thread, void *status) is called to block the calling thread until the specified thread thread is finished. It is a synchronization routine, as its name suggests. All Pthreads are joinable by default.
- int pthread_cancel(pthread_t thread) is a called to cancel the thread thread.

Note that the main() function of a program comprises a single default thread. All other threads must be explicitly created by calling pthread_create(). Threads can be created from anywhere of your code. Once created, threads are peers and may create other threads.

The code listed in Figure 6.1 shows how to create a thread with default attributes and a thread with customized attributes. Line 1 includes pthread.h header file, because all Pthread APIs are defined in the file. Line 4 declares a thread attribute object, which is initialized in Line 41 by calling the pthread_attr_init() routine. Line 46 sets the detachstate attribute to PTHREAD_CREATE_JOINABLE by calling the pthread_attr_getdetachstate() routine.

In the main function starting from Line 36, two thread IDs, namely threadD and threadC, are declared in Line 37. Line 51 creates a thread with the default attribute object by setting the attribute object reference to a NULL. The routine of this thread is threadDefault() in Lines 6–10. Line 57 creates a thread with the constructed attribute object attr. The routine of this thread is threadCustomized() in Lines 12–34. Lines 61 and 62 call the pthread_join() routine to synchronize the two created threads with the default main thread. Line 65 frees up the memory of the attribute object. Line 66 exits the main thread and thus the entire program.

The thread created with the default attribute object simply outputs a message to show that the thread is created and running. In the routine of the thread created with the customized attribute object, it retrieves the detach state from the attribute object attr in Line 18 by calling the routine pthread_attr_getdetachstate() and then prints it out. It also retrieves the scheduling policy from attr in Line 25. Since we didn't set up the policy value before the thread was created, the value retrieved is a default one.

```
1.  #include <pthread.h>
2.  #include <stdio.h>
3.
4.  pthread_attr_t attr;
5.
6.  void *threadDefault(void *arg) {
7.      printf("A thread with default attributes is created!\n\n");
8.      pthread_exit(NULL);
9.      return NULL;
10. }
11.
12. void *threadCustomized(void *arg) {
13.     int policy;
14.     int detachstate;
15.     printf("A thread with customized attributes is created!\n");
16.
17.     /* Print out detach state */
18.     pthread_attr_getdetachstate(&attr, &detachstate);
19.     printf("  Detach state: %s\n",
20.         (detachstate == PTHREAD_CREATE_DETACHED) ?
                "PTHREAD_CREATE_DETACHED" :
21.         (detachstate == PTHREAD_CREATE_JOINABLE) ?
                "PTHREAD_CREATE_JOINABLE" :
22.         "???");
23.
24.     /* Print out scheduling policy */
25.     pthread_attr_getschedpolicy(&attr, &policy);
26.     printf("  Scheduling policy: %s\n\n",
27.         (policy == SCHED_OTHER) ? "SCHED_OTHER" :
28.         (policy == SCHED_FIFO)  ? "SCHED_FIFO" :
29.         (policy == SCHED_RR)    ? "SCHED_RR" :
30.         "???");
31.
32.     pthread_exit(NULL);
33.     return NULL;
34. }
35.
36. int main(int argc, char* argv[]) {
37.     pthread_t threadD, threadC;
38.     int rc;
39.
40.     /* Inlitialize attributes */
41.     rc = pthread_attr_init(&attr);
42.     if (rc)
43.         printf("ERROR; RC from pthread attr init() is %d \n", rc);
44.
45.     /* Set detach state and  */
```

Figure 6.1 Pthreads creation and termination.

```
46.       rc = pthread_attr_setdetachstate(&attr,
            PTHREAD_CREATE_JOINABLE);
47.       if (rc)
48.          printf("ERROR; RC from pthread_attr_setdetachstate()
               is %d \n", rc);
49.
50.       /* Creating thread  with default attributes */
51.       rc = pthread_create(&threadD, NULL, threadDefault, NULL);
52.       if (rc)
53.          printf("ERROR when creating default thread; Code is
               %d\n", rc);
54.
55.       /* Creating thread  with constructed attribute object */
56.       rc = pthread_create(&threadC, &attr, threadCustomized,
            NULL);
57.       if (rc)
58.          printf("ERROR when creating customized thread;
               Code is %d\n", rc);
59.
60.       /* Synchronize all threads */
61.       pthread_join(threadD, NULL);
62.       pthread_join(threadC, NULL);
63.
64.       /* Free up attribute object and exit */
65.       pthread_attr_destroy(&attr);
66.       pthread_exit(NULL);
67. }
```

Figure 6.1 *(Continued)*

Note that Lines 9 and 33 will never get hit; they are there to *cheat* the compiler.

The output of this program is shown in Figure 6.2.

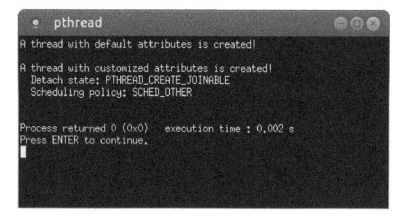

Figure 6.2 Output of the program listed in Figure 6.1.

6.3   Synchronization Primitives

Concurrent programs should be constructed carefully so that processes or threads can exchange information, but do not interfere with each other and cause errors. They often need to satisfy certain constraints on the interleaving of their executions. This kind of synchronization requirement is necessary due to race condition behavior. To prevent multiple tasks from accessing shared data effectively, most RTOS kernels provide synchronization objects to help programmers in implementing synchronization. Among them, mutexes, condition variables, and semaphores are the most popular kernel objects. This section first explains the concepts of race conditions and critical sections and then introduces the various solutions to process synchronization.

6.3.1   Race Conditions and Critical Sections

In concurrent computing, there are multiple processes or threads running concurrently. Generally, programmers have no control over when processes are swapped. Swapping or preemption occurs under the control of the scheduler of the operating system, not the programmer. Each process might get interrupted after any instruction. A *race condition* occurs when two or more processes interact via shared data, and the final outcome depends on the exact instruction sequence of the processes.

Consider an ATM application that has two processes: a process P1 that credits money to an account and a process P2 that debits money from an account. Each process performs the following three actions sequentially:

- Reads the balance of an account
- Modifies the balance
- Writes the balance back

Suppose that the balance of a joint account of a couple is $1000. While the husband is depositing $200 to the account at an ATM, the wife is withdrawing $200 from the account at another ATM. If the two processes run according to the following sequence:

- P1: Read the balance of the account
- P1: Increase the balance
- P1: Write the balance back
- P2: Read the balance of the account
- P2: Decrease the balance
- P2: Write the balance back

or the following sequence:

- P2: Read the balance of the account
- P2: Decrease the balance

- P2: Write the balance back
- P1: Read the balance of the account
- P1: Increase the balance
- P1: Write the balance back

The final balance will remain $1000, which is correct. However, if the two processes interleave as follows:

- P1: Read the balance of the account
- P2: Read the balance of the account
- P1: Increase the balance
- P1: Write the balance back
- P2: Decrease the balance
- P2: Write the balance back

The final balance will be $800, which is incorrect. The root cause to the issue is that here the deposit operations are not *mutually exclusive*. Mutually exclusive operations are those that cannot be interrupted. An example is accessing a shared memory location.

In a program, a code segment that accesses shared data is called a *critical section*. Being inside a critical section is a special status of a task. When one task enters its critical section to access the shared memory, all other tasks needing access to that shared data should wait. This way race conditions are avoided. Because in a real-time system, tasks need to meet their deadlines, critical sections must be coded as short as possible. Normally, a task performs only noncritical work on local variables the majority of the time. We should not place unnecessary code in a critical section. Moreover, coding errors in a critical section such as an infinite loop should also be carefully checked and removed.

Chapter 5 mentioned that before a task enters a critical section, it needs to lock a mutex or semaphore. We introduce this kind of task synchronization primitives offered by most RTOSs in the next few sections.

6.3.2 Mutex

Mutex is the short form for *mutual exclusion object*. It is used to allow multiple tasks to share the same resource, such as global data, but not simultaneously. The two basic operations on a mutex object are as follows:

- LOCK(*mutex*) It blocks the calling task until the object *mutex* is available and then makes it unavailable to other tasks.
- UNLOCK(*mutex*) It unlocks the object *mutex* and makes it available to other tasks.

In a program in which resources will be shared by multiple tasks, a mutex object is created with a unique name for each resource. After that, any task that needs a shared resource must lock the corresponding mutex object from other tasks while it is using the resource. The mutex object is set to unlock when the data is no longer needed or the task is finished.

There are three routines dealing with mutex operations in POSIX:

- `int pthread_mutex_lock(pthread_mutex_t *mutex)` This routine locks the mutex object referenced by `mutex`. If the object is already locked by another thread, this call will block the calling thread until it is unlocked.
- `int pthread_mutex_unlock(pthread_mutex_t *mutex)` This routine unlocks the mutex object referenced by `mutex` if called by the owning thread. If it is called by a routine that did not lock the mutex, or the mutex was already unlocked, an error is returned.
- `int pthread_mutex_trylock(pthread_mutex_t *mutex)` This routine is called to lock the mutex object referenced by `mutex`. However, if the mutex is already locked, it retunes immediately with a "busy" error code.

Each of these function calls returns zero upon success; otherwise, an error number that indicates the error type is returned. The type of mutex objects is `pthread_mutex_t`. Mutex objects must be initialized through the function call of `pthread_mutex_init()` before they are used.

The code listed in Figure 6.3 illustrates the use of mutex objects. This program creates two threads that are identified as `thread1` and `thread2`. The routine that implements thread1 is `deposit()`, which prompts user to enter a double number and then adds it to the global variable `balance`. Before the routine enters its critical section to change the global variable, the `pthread_mutex_lock()` routine is called in Line 11 to lock the global mutex object `my_mutex`, which is declared in Line 5 and initialized in Line 50. After the access to `balance` is established, the `pthread_mutex_unlock()` routine is called in Line 17 to unlock the mutex object. Notice that in general the code in Lines 13, 14, and 16 should not be included in a critical section, because they will affect its running time. As an illustrative example, we place them inside the critical section only for the convenience of experiment; otherwise, the output of the program will be messy. The routine that implements `thread2` is `withdraw()`, which prompts users to enter a double number and then subtract it from `balance`.

In Line 61, the `pthread_mutex_destroy()` routine is called to destroy the mutex object and free up the memory. It is very important to notice that before the mutex object is destroyed, we need to make sure that the two threads have completed accessing it. To ensure that, in Lines 57 and 58, the `pthread_join()` routine is called to block the default main routine from moving on until the two threads are done, that is, the `pthread_exit()` routine is called in Lines 19 and 42.

```
1.  #include "pthread.h"
2.  #include <stdio.h>
3.
4.  double balance=0;
5.  pthread_mutex_t my_mutex;
6.
7.  void *deposit(void *dummy){
8.      double credit = 0;
9.
10.     /* enter critical section */
11.     pthread_mutex_lock(&my_mutex);  /* lock mutex */
12.         /* put printf and scanf inside critical section ONLY for
               experiment */
13.         printf("\nI am in thread 1. Enter amount to deposit: ");
14.         scanf("%lf", &credit);
15.         balance = balance + credit;
16.         printf("The new balance is: %lf\n", balance);
17.     pthread_mutex_unlock(&my_mutex); /* unlock mutex */
18.
19.     pthread_exit(NULL);
20.     return NULL;
21. }
22.
23. void *withdraw(void *dummy){
24.     double debit = 0;
25.
26.     /* enter critical section */
27.     pthread_mutex_lock(&my_mutex);   //lock mutex
28.         /* put printf and scanf inside critical section ONLY for
               experiment */
29.         printf("\nI am in thread 2. Enter amount to withdraw: ");
30.         while (1){
31.             scanf("%lf", &debit);
32.             if (balance - debit < 0)
33.                 printf("Insufficient balance. Please enter a
                       smaller amount: ");
34.             else
35.                 break;
36.         }
37.
38.         balance = balance - debit;
39.         printf("The new balance is: %lf\n", balance);
40.     pthread_mutex_unlock(&my_mutex);   //unlock mutex
41.
42.     pthread_exit(NULL);
43.     return NULL;
44. }
45.
46. int main(int argc, char* argv[]){
```

Figure 6.3 Use of mutex objects.

```
47.      pthread_t thread1, thread2;
48.
49.      /* initialize mutex */
50.      pthread_mutex_init(&my_mutex, NULL);
51.
52.      /* creating threads */
53.      pthread_create(&thread1, NULL, deposit, NULL);
54.      pthread_create(&thread2, NULL, withdraw, NULL);
55.
56.      /* wait until threads to finish */
57.      pthread_join(thread1, NULL);
58.      pthread_join(thread2, NULL);
59.
60.      /* delete mutex */
61.      pthread_mutex_destroy(&my_mutex);
62.
63.      pthread_exit(NULL);
64. }
65.
```

Figure 6.3 (Continued)

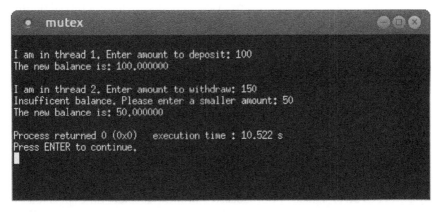

Figure 6.4 Execution of the program listed in Figure 6.3.

Figure 6.4 shows a snapshot of the execution of the program. Keep in mind that you never know which of the two created threads executes first, even if we call the creation routine for thread1 before thread2. In other words, when you execute the program, you may see that the "I am in thread2. Enter amount to withdraw:" message appears first.

6.3.3 Condition Variables

In some cases, as a task executes in a critical section, it discovers that it cannot proceed until some other task takes a particular action on the protected data. *Condition variables* are synchronization objects used in conjunction with mutex objects to allow tasks to synchronize based upon the actual value of some data. Condition variables should be global to all tasks that may manipulate them. Two fundamental operations on condition variables are as follows:

- WAIT(*condition, mutex*). Suspends the invoking task until another task performs SIGNAL(*condition*).
- SIGNAL(*condition*). Resumes exactly one other task that is suspended due to WAIT operation. If no task is waiting, then the signal will have no effect.

Because a condition variable is a global variable, access to it has to be protected by a mutex object.

In POSIX, the WAIT operation is carried out by calling the pthread_cond_wait() routine:

```
int pthread_cond_wait(pthread_cond_t *condition,
        pthread_mutex_t *mutex);
```

It blocks the calling thread until the condition object referenced by the first argument condition is signaled. While it waits, the mutex object referenced by the second argument mutex is released and thus unblocks a thread on the mutex. When the condition variable is signaled, the calling thread is awakened, and the mutex object is locked for use by the calling thread. Upon successful completion, a value of zero shall be returned; otherwise, an error number shall be returned to indicate the error.

A similar routine to pthread_cond_wait() is

```
int pthread_cond_timewait(pthread_cond_t *condition,
        pthread_mutex_t *mutex,
        const struct timespec *restrict abstime);
```

The only difference is that this time wait routine shall return an error if the absolute time specified by abstime passes before the condition referenced by condition is signaled or broadcasted or if the absolute time specified by abstime has already been passed at the time of the call. In real-time applications, this routine is more helpful in implementing real-time tasks.

There are two POSIX routines that can be used to signal a condition:

- int pthread_cond_signal(pthread_cond_t *condition)
  This routine signals a thread that is waiting on the specified condition variable. If no thread is waiting, then it takes no effect.
- int pthread_cond_broadcast(pthread_cond_t *condition) This routine signals all threads that are waiting on the specified condition object. If no thread is waiting, then it takes no effect. Otherwise, when the threads that were the target of the broadcast wake up, they contend for the mutex object that they have been associated with the condition variable on the call to pthread_cond_wait().

To illustrate how condition variables are used to solve task synchronization issues, we consider the classic *producer–consumer problem*, which is also known as the *bounded-buffer problem*. In this problem, producers and consumers share a common, fixed-size buffer as a queue. The producers' job is to repeatedly generate a piece of data and add it into the buffer. At the same time, the consumers are consuming the data, that is, removing it from the buffer one piece at a time. No producer should try to add data into the buffer if it is full. Similarly, no consumer should try to remove data from an empty buffer.

The solution is that if a producer finds the buffer is full, it stays waiting until it is notified by a consumer that an item is removed from the buffer, and then, the producer starts to fill the buffer. In the same way, the consumer can go to sleep if it finds the buffer to be empty. The next time the producer adds data into the buffer, it wakes up the sleeping consumer.

To solve this problem, we need a mutex object to protect the shared buffer so that it never gets accessed by multiple tasks concurrently. We also need a global count to track the number of items in the buffer. If it is equal to 0, then consumer tasks should be blocked; if it is equal to the buffer size, then producer tasks should be blocked. Executing a producer task may wake up a consumer task, while executing a consumer task may unblock a producer task. The blocking and unblocking can be achieved by using a condition variable object.

Figure 6.5 lists a program that solves the producer–consumer problem. The program implements two producers and one consumer. The buffer is implemented by an array of integers with a capacity of 4. The number of integers in the array is tracked by the global variable size. Lines 85 and 86 indicate that the two producer threads share the same routine produce(). The consumer thread is implement by the consume() routine. A mutex object and a condition variable object are declared in Lines 9 and 10, respectively. Each producer thread iterates five times and thus writes five numbers to the buffer. Each time a producer thread writes a number to the buffer, it sleeps for 1 millisecond, as shown in Line 38, to allow other threads to run. The consumer thread iterates 10 times and thus reads and removes 10 numbers from the buffer. Each time the consumer thread removes a number from the buffer, it sleeps for 2 milliseconds, as shown in Line 69.

Let us see how the mutex and condition objects are used together to solve the synchronization issue in the program. After a producer thread locks the mutex object in Line 17, the if-statement in Lines 19–24 checks to see if the buffer is full. If it is full, the call of pthread_cond_wait(&my_CV, &my_mutex) blocks the thread on the condition variable my_CV until it is signaled. If the buffer is not full, or it was full but a space was just freed up (my_CV was signaled), it continues by adding an integer in the buffer (Line 26), incrementing the size of the buffer (Line 28). If after the increment, the size is equal to 1, then the buffer was empty before, which means that the consumer thread might be blocked. Therefore, in Line 32, the pthread_cond_signal() routine is

```
1.  #include <pthread.h>
2.  #include <stdio.h>
3.  #include <stdlib.h>
4.  #include <unistd.h>
5.
6.  #define BUFFER_SIZE 4
7.
8.  int theArray[BUFFER_SIZE], size=0;
9.  pthread_mutex_t myMutex;
10. pthread_cond_t myCV;   /* declare condition variable object */
11.
12. void *produce(void *arg) {
13.     int id = (int)arg;
14.     int i;
15.
16.     for (i = 0; i<5; i++){
17.         pthread_mutex_lock(&myMutex);   /* lock mutex */
18.
19.         if (size == BUFFER_SIZE){
20.             printf("Producer %d waiting...\n", id);
21.
22.             /* wait for a space to be freed up */
23.             pthread_cond_wait(&myCV, &myMutex);
24.         }
25.
26.         theArray[size] = i;
27.         printf("Producer %d added %d.\n", id, i);
28.         size++;
29.
30.         if (size == 1) {
31.             /* signal consumer to resume */
32.             pthread_cond_signal(&myCV);
33.         }
34.
35.         pthread_mutex_unlock(&myMutex);   /* unlock mutex */
36.
37.         /* sleep for 1 millisecond, so that other threads can run */
38.         usleep(1000);
39.     }
40.     return NULL;
41. }
42.
43. void *consume(void *arg) {
44.     int item;
45.     int i;
46.
47.     for (i=0; i<10; i++){
48.         pthread_mutex_lock(&myMutex);   /* lock mutex */
49.
50.         if (size == 0) {
51.             printf("Consumer waiting...\n");
52.
53.             /* waiting for an item to be added */
```

Figure 6.5 Solving the producer–consumer problem with condition variables.

```
54.                pthread_cond_wait(&myCV, &myMutex);
55.            }
56.
57.            item = theArray[size-1];
58.            printf("Consumer removed %d.\n", item);
59.            size--;
60.
61.            if (size == BUFFER_SIZE-1) {
62.                /* signal producer to resume */
63.                pthread_cond_signal(&myCV);
64.            }
65.
66.            pthread_mutex_unlock(&myMutex);  /* unlock mutex */
67.
68.            /* sleep for 1 millisecond so that other threads can run */
69.            usleep(1000);
70.        }
71.
72.    return NULL;
73. }
74.
75. int main(int argc, char* argv[]) {
76.     int t1=1, t2=2;
77.     pthread_t consumer, producer1, producer2;
78.
79.     /* Initialize mutex and condition variable objects */
80.     pthread_mutex_init(&myMutex, NULL);
81.     pthread_cond_init (&myCV, NULL);
82.
83.     /* Create one consumer thread and two producer threads */
84.     pthread_create(&consumer, NULL, consume, NULL);
85.     pthread_create(&producer1, NULL, produce, (void *)t1);
86.     pthread_create(&producer2, NULL, produce, (void *)t2);
87.
88.     /* Wait for all threads to complete */
89.     pthread_join(consumer, NULL);
90.     pthread_join(producer1, NULL);
91.     pthread_join(producer2, NULL);
92.
93.     /* Clean up and exit */
94.     pthread_mutex_destroy(&myMutex);
95.     pthread_cond_destroy(&myCV);
96.     pthread_exit(NULL);
97.
98.     return 0;
99. }
```

Figure 6.5 (*Continued*)

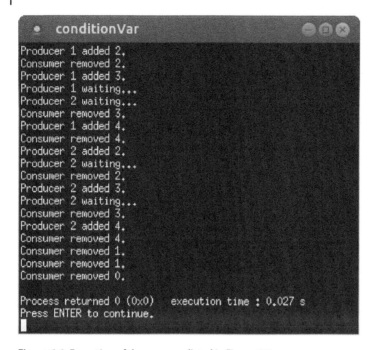

Figure 6.6 Execution of the program listed in Figure 6.5.

called to signal the consumer thread to resume its execution. Line 35 calls the pthread_mutex_unlock() routine to exit the critical section.

The behavior of the consumer thread is identical to that of the producer threads. It locks the mutex in Line 48 first, and then, the if-statement in Lines 50–55 checks to see if the buffer is empty. If it is empty, the pthread_cond_wait() routine is called to block the thread until the condition variable my_CV is signaled. It can only be signaled by a producer thread that executes the statement in Line 32. If the buffer is not empty, or it was empty but an integer was just added (my_CV was signaled), it continues by taking an integer out of the buffer (Lines 57 and 59). The if-statement in Lines 61–64 checks to see if the buffer was full before an item was removed. If it was full, one or two producer threads might be blocked. Therefore, in Line 63, the pthread_cond_signal() routine is called to wake up one producer thread.

Figure 6.6 shows a snapshot of the program execution result.

6.3.4 Semaphores

A semaphore is a synchronization primitive designed by Edsger Dijkstra in the 1960s. A semaphore is a *shared counter* in essence, which is associated with two operations:

Table 6.2 Operations on a semaphore initialized to 1.

Action #	Task	Operation	Value	Consequence
1	A	P()	0	Task A moves on
2	A	P()	0	Task A is blocked
3	B	V()	0	Task A is unblocked
4	A	V()	1	—
5	B	P()	0	Task B moves on
6	A	V()	1	—
7	B	V()	2	—

- P(*sem*) It is an atomic action that waits for the value of the semaphore *sem* to be greater than 0 and then decrements it.
- V(*sem*) It is an atomic action that increments the value of the semaphore *sem* by 1.

Table 6.2 showcases the behavior of semaphores.

Notice that after the third action, the semaphore value is increased to 1 first and then decremented to 0 immediately by the call of P() from task A.

The initial value of a semaphore has a great impact on the consequences of semaphore operations. If we change the initial value to 0 or 2, the results listed in Table 6.2 will be completely different. Specifically, if the initial semaphore value is 0, then task A is blocked upon Action #1; if the initial semaphore value is 2, then task A can move on after both Action #1 and Action #2.

Semaphores can be used to implement blocks. For example, the following pseudocode shows that a semaphore is used to protect the access to the balance of a bank account:

```
semaphore mySem = 1;

void deposit(account, amount){
        P(mySem);
                balance = get_balance(account);
                balance -= amount;
                put balance(account, balance);
        V(mySem);
}
```

Although a semaphore can implement the functions of a mutex, there is a big difference between them. A mutex object has an owner, but a semaphore object does not. When a mutex object is locked by a task, it can only be unlocked by the same task. On the other hand, when P(mySem) is called in one task and blocks the calling task, a subsequent V(mySem) can be called in any other task to unblock the first task.

As an example, we look into the classic readers–writers problem and discuss its semaphore-based solution. The problem is stated as follows: many tasks try to access the same shared resource at one time. Some tasks may read and some may write. The constraint is that when a task is in the act of writing to the shared data, no other task may access it for either reading or writing. However, it *is* allowed for two or more readers to access the shared data at the same time.

A solution to the problem can be illustrated with the following pseudocode:

```
semaphore write = 1;
int readcount = 0;

reader() {
    readcount++;
    if (readcount == 1) {
        P(write);
    }
    do_read(); // access shared data
    readcount--;
    if (readcount == 0) {
        V(write);
    }
}

writer() {
    P(write);
        do_write(); // access shared data
    V(write);
}
```

Logically, this solution seems to be okay: a reader task is allowed to read the shared data only if currently there are tasks reading the shared data (the value of readcount is greater than 1), and if there is no task currently reading the shared data, the reader task calls P() to lock the semaphore. On the writer side, a writer task always calls P() to lock the semaphore before it can write to the shared data. However, since there are multiple reader and writer tasks in the system and context switches can take place at any moment, the following situation may occur:

- Reader task #1 executes till the completion of readcount++;
- Reader task #1 swaps out;

- Reader task #2 swaps in and executes `readcount++`;
- Reader task #2 swaps out;
- Writer task #1 swaps in, executes P() and starts running `do_write();`
- Writer task #1 swaps out;
- Reader task #1 swaps in, skips P() because now the value of readcount is 2, and starts reading.

So, the system enters a state while a writer task is writing, a reader task is reading. This is an error. The root cause to the error is that multiple tasks make change to readcount before the test and P() are completed. It is a race condition issue. To correct the error, we can introduce another semaphore to make "increment, test, P()" and "decrement, test, V()" both atomic. The pseudocode with the fix is listed as follows:

```
semaphore read = 1;
semaphore write = 1;
int readcount = 0;

reader() {
    P(read);
    readcount++;
    if (readcount == 1) {
        P(write);
    }
    V(read)
    do_read(); // access shared data
    P(read)
    readcount-;
    if (readcount == 0) {
        V(write);
    }
    V(read);
}

writer() {
    P(write);
        do write(); // access shared data
    V(write);
}
```

In POSIX, semaphores are not included in the Pthreads library. The necessary declarations related to semaphores are contained in semaphore.h. The semaphore routines include the following:

- `int sem_wait(sem_t *sem)`: This is the P() operation on the semaphore object referenced by sem.
- `int sem_post(sem_t *sem)`: This is the V() operation.
- `int sem_init(sem_t *sem, int pshared, unsigned int val)`: Initialize a new semaphore object referenced by sem to the value val. Note that the second argument denotes *how* the semaphore will be shared. Passing zero denotes that it will be shared among threads rather than processes.
- `int sem_destroy(sem_t *sem)`: Deallocate the semaphore object referenced by sem.

The code listed in Figure 6.7 implements the solution of the readers–writers problem that we discussed before. There are a few points worth mentioning here: first, the routines of the reader threads and writer threads are implemented as an infinite while-loop. This is a typical way to implementing a periodic task. To stop the running of these threads, the `pthread_cancel()` routine is called in the main function. Second, the focus of this implementation is on thread synchronization, and thus, the code of reading and writing the shared data is omitted. Third, the output of this program varies with the four constants defined in Lines 7–10.

```
1.  #include <semaphore.h>
2.  #include <pthread.h>
3.  #include <stdio.h>
4.  #include <stdlib.h>
5.  #include <unistd.h>
6.
7.  #define READERS 5
8.  #define WRITERS 3
9.  #define READER_SLEEP_TIME 20000
10. #define WRITER_SLEEP_TIME 50000
11. #define MAIN_SLEEP_TIME 5000000
12.
13. sem_t semRead;
14. sem_t semWrite;
15. int readCount;
16.
17. void *reader(void *arg) {
18.     int *p = (int *)arg;
19.
20.     while(1) {
21.         /* lock semaphore semRead to update readCount*/
22.         sem_wait(&semRead);
23.         readCount++;
24.         printf("                    Number of readers: %d
            \n", readCount);
25.
```

Figure 6.7 Solving the readers–writers problem with semaphores.

```
26.              if (readCount == 1) {
27.                  /* lock semaphore semWrite */
28.                  sem_wait(&semWrite);
29.              }
30.
31.              /* release semaphore semRead */
32.              sem_post(&semRead);
33.
34.              /* entered critical section. reading code goes here */
35.              printf("                    Reader #%d reading
                     ...\n", (int) *p);
36.
37.              /* lock semaphore semRead to update readCount*/
38.              sem_wait(&semRead);
39.              readCount-;
40.
41.              if (readCount == 0) {
42.                  /* release semaphore semWrite */
43.                  sem_post(&semWrite);
44.              }
45.
46.              /* release semaphore semRead */
47.              sem_post(&semRead);
48.              usleep(READER_SLEEP_TIME);
49.          }
50.          return NULL;
51. }
52.
53. void *writer(void *arg) {
54.     int *p = (int *)arg;
55.     while(1){
56.          sem_wait(&semWrite);
57.
58.          /* writing code goes here */
59.          printf("Writer #%d writing ...\n", (int) *p);
60.
61.          sem_post(&semWrite);
62.          usleep(WRITER_SLEEP_TIME);
63.     }
64.     return NULL;
65. }
66.
67. int main(int argc, char* argv[]) {
68.     pthread_t readers[READERS];
69.     pthread_t writers[WRITERS];
70.     int i, rc, r[READERS], w[WRITERS];
71.     readCount = 0;
72.
73.     /* initialize the semaphores to 1. */
74.     sem_init(&semRead, 0, 1);
```

Figure 6.7 (*Continued*)

```
75.     sem_init(&semWrite, 0, 1);
76.
77.     for(i = 0; i < WRITERS; i++){
78.         w[i] = i;
79.         rc = pthread_create(&writers[i], NULL, writer, (void *)&w[i]);
80.         if (rc){
81.             perror("In writer pthread_create()");
82.             exit(1);
83.         }
84.         usleep(20000);
85.     }
86.
87.     /* create reader and writer threads */
88.     for(i = 0; i < READERS; i++){
89.         r[i] = i;
90.         rc = pthread_create(&readers[i], NULL, reader, (void *)&r[i]);
91.         if (rc){
92.             perror("In reader pthread_create()");
93.             exit(1);
94.         }
95.         usleep(20000);
96.     }
97.
98.     usleep(MAIN_SLEEP_TIME);
99.
100.    /* cancel all threads */
101.    for(i = 0; i < WRITERS; ++i)
102.        pthread_cancel(writers[i]);
103.    for(i = 0; i < READERS; ++i)
104.        pthread_cancel(readers[i]);
105.
106.    /* destroy semaphores */
107.    sem_destroy(&semRead);
108.    sem_destroy(&semWrite);
109.
110.    return 0;
111.}
```

Figure 6.7 (*Continued*)

A snapshot of the output of the program is shown in Figure 6.8.

6.4   Communication among Tasks

Operating systems provide intertask communication mechanisms for tasks to share data. Message queues, pipes, named pipes, sockets, and shared memory are among the most popular mechanisms being used. We introduce message queues and shared memory in this section.

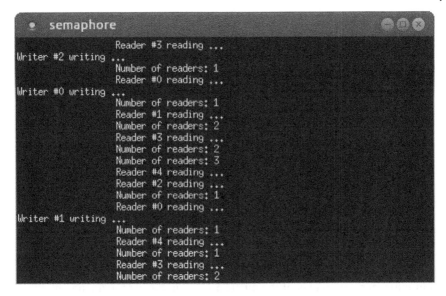

Figure 6.8 A snapshot of the output of the program listed in Figure 6.7.

6.4.1 Message Queues

A message queue is used for intertask information transfer. Two basic operations are as follows:

- SEND(*msg*) It sends the message *msg*.
- RECEIVE(*msg*) It receives a message and stores it in *msg*.

Message contents can be anything that is mutually comprehensible between senders and receivers. Typically, a message is an instance of a data structure.

In *direct message passing*, each process wanting to communicate must *explicitly* name the recipient or sender of the communication. Therefore, the two operations must have the counterpart of the communication as an argument:

- SEND(*P*, *msg*) It sends the message *msg* to the task *P*.
- RECEIVE(*Q*, *msg*) It receives a message from the task *Q* and stores it in *msg*.

In *indirect message passing*, messages are sent to and received from shared message queues, mailboxes, or ports. A message queue can be viewed as an object into which messages are placed by tasks and from which messages can be removed by other tasks. In a real application, there might be multiple message queues. The operation model of message queues in indirect message passing is as follows:

- SEND(*MQ*, *msg*) It sends the message *msg* to the message queue *MQ*.
- RECEIVE(*MQ*, *msg*) It receives a message from the message queue *MQ* and stores it in *msg*.

Communication with message queues can proceed *synchronously* or *asynchronously*. In synchronous communication, the sending task blocks itself until the message it sends is received by the receiver. In asynchronous communication, the sender can continue with other processing even if the receiver has not yet received the message. Of course, in asynchronous communication, a message buffering mechanism needs to be in place to store the message until the receiver retrieves the message. In general, asynchronous communication is more desired because it increases the level of concurrency. Communication can be one-to-one, many-to-one, one-to-many, and many-to-many.

Before a task can send or receive a message, a message queue must be created and initialized. Attributes of the queue are set up during the queue initialization. The attributes include the maximum size of the queue, the maximum length of a message, the type of communication (blocking or nonblocking), and so on.

In POSIX message queues, messages are queued (and thus received) in the order of priority. The message attribute structure is defined as follows:

```
struct mq_attr {
    long int mq_flags;   /* Message queue flags. */
    long int mq_maxmsg;  /* Maximum # of messages. */
    long int mq_msgsize; /* Maximum message size. */
    long int mq_curmsgs; /* # of messages in queue. */
};
```

Message-queue-related definitions are included in the mqueue.h head file. The routine that creates a new or opens an existing POSIX message queue is

```
mqd_t mq_open(const char *name,
        int oflag,
        mode_t mode,
        struct mq_attr *attr
);
```

It creates and opens a queue referenced by name for access. The message queue name must follow the construction rules as for a normal file path name. In particular, it has to start with a "/," and the name cannot contain additional "/."

The second argument `oflags` controls the way in which the message queue is opened. It can be `O_RDONLY` for receiving messages, `O_WRONLY` for sending messages, or `O_RDWR` for both sending and receiving operations on the queue. The flag can be OR'ed with `O_CREATE`, meaning that the routine is called to create a queue. Only when `O_CREATE` is used, the last two arguments are necessary. It may also be OR'ed with other flags. For example, you can specify `O_NONBLOCK` to use the queue in a nonblocking mode. By default, `mq_send()` would block if the queue is full and `mq_receive()` would block if the queue is empty. But if `O_NONBLOCK` is specified in `flag`, the call would return in those cases immediately with an error. Therefore, a flag

```
O_RDONLY | O_CREATE
```

would mean that the routine is called to create a new queue for read-only, and when the queue is empty, `mq_receive()` call will be blocked until a message is sent to the queue.

The `mode` argument is a bit mask that specifies the access permission to be placed on the queue. The bit values that may be specified are the same as those for files. For example, 0222 means write-only, 0444 means read-only, and 0666 means read and write.

The last argument `attr` is a reference to the attribute structure instance associated with the queue. If it is NULL, then the queue is created with the implementation-defined default attributes.

The `mq_open()` call returns the descriptor of the message queue if the queue is created or opened successfully. The descriptor is used to reference the message queue by most other message queue routines. The call fails if the queue reference already exists.

To open an existing queue, we call

```
mqd_t mq_open(const char *name,
              int oflag);
```

which is the same routine but with two arguments only. In this case, `O_CREATE` should not appear in the `oflag` argument.

The routine that sends messages is

```
int mq_send(mqd_t mqdes,
            const char* msgbuf,
            size_t length,
            unsigned int priority);
```

It adds the message pointed by `msgbuf` to the message queue referenced by the descriptor `mqdes`. The `length` must be less than or equal to the `mq_msgsize` specified in the attributes with which the queue is created. The last argument specifies the priority of the message. Messages of the same priority are stored on the queue in FIFO order.

The routine that receives messages is

```
size_t mq_receive(mqd_t mqdes,
                  char *msgbuf,
                  size_t length,
                  unsigned int *priority);
```

It retrieves a message from the queue referenced by the descriptor mqdes. The retrieved message is removed from the queue and stored in the area pointed by msgbuf, whose length is length. Messages are retrieved from the queue in FIFO order within priorities. Messages of higher priorities are retrieved first. The priority of the message is stored in priority. Upon success, the call returns the number of bytes in the received message. Otherwise, it returns −1.

Other important routines include the following:

- int mq_setattr(mqd_t mqdes, struct mq_attr *new_attr, struct mq_attr *old_attr) It sets some attributes of the message queue referenced by the descriptor mqdes. New attributes are set from the values given in the structure referenced by new_attr. The old attributes are stored in the location referenced by old_attr, if the pointer is not NULL. However, the only attribute that can be modified by this function call is the O_NONBLOCK flag in mq_flags. Other fields in the structure pointed by new_attr are ignored.
- int mq_getattr(mqd_t mqdes, struct mq_attr * attr) It retrieves the attributes of the message queue referenced by the descriptor mqdes and stores it in the buffer pointed by attr.
- int mq_close(mqd_t mqdes) It terminates access to a message queue referenced by the descriptor mqdes.
- int mq_unlink(const char *name) It removes the name of the message queue referenced by name and marks the queue for deletion when all processes have closed it.

Messages can be sent or received with a time-out. That is, in case the flag NONBLOCK is not set, the message queue call, be it sending messages or receiving messages, will block for a time limited by the specified argument. The sending and receiving routines with time-out are

- mq_timedsend(mqdes, msgbuf, length, priority, timeout);
- mq_timedreceive(mqdes, msgbuf, length, priority, time-out);

where time-out is an absolute value in seconds and nanoseconds since Epoch. To use a relative time-out, one way is to call the clock_getting() routine to get the current time and then add the relative amount of time.

Figure 6.9 lists a program (*sender*) that creates a message queue and then sends a message to the queue, while Figure 6.10 lists a program (*receiver*) that

```
1.  #include <stdio.h>
2.  #include <mqueue.h>
3.  #include <sys/stat.h>
4.  #include <stdlib.h>
5.  #include <unistd.h>
6.  #include <string.h>
7.  #include <errno.h>
8.  #define QUEUE_NAME     "/my_queue"
9.  #define MAX_MSG_LEN    100
10.
11. int main(int argc, char *argv[]) {
12.     mqd_t myQ_id;
13.     unsigned int msg_priority = 0;
14.     pid_t my_pid = getpid();
15.     char msgcontent[MAX_MSG_LEN];
16.
17.     /* create a message queue */
18.     myQ_id = mq_open(QUEUE_NAME, O_RDWR | O_CREAT | O_EXCL,
19.                                  S_IRWXU | S_IRWXG, NULL);
20.
21.     /* if not successful */
22.     if (myQ_id == (mqd_t)-1) {
23.         /* if the queue already exists, simply open it */
24.         if (errno == EEXIST){
25.             myQ_id = mq_open(QUEUE_NAME, O_RDWR );
26.                   if (myQ_id == (mqd_t)-1) {
27.                         perror("In mq_open(2)");
28.                         exit(1);
29.                   }
30.         }
31.         else {
32.             perror("In mq_open(4)");
33.             exit(1);
34.         }
35.     }
36.
37.     /* compose a message */
38.     snprintf(msgcontent, MAX_MSG_LEN, "Hello from process
        %u", my_pid);
39.
40.     /* send the message */
41.     if (mq_send(myQ_id, msgcontent, strlen(msgcontent)+1,
        msg_priority) == 0){
42.         printf("A message is sent. \n");
43.         printf("   Content: %s\n", msgcontent);
44.     }
45.     else {
46.             perror("In mq_send()");
47.             exit(1);
48.     }
49.
50.     /* close the queue */
51.     mq_close(myQ_id);
52.
53.     return 0;
54. }
```

Figure 6.9 Sending messages to a message queue.

```
1.  #include <stdio.h>
2.  #include <mqueue.h>
3.  #include <sys/stat.h>
4.  #include <stdio.h>
5.  #include <mqueue.h>
6.  #include <stdlib.h>
*7. #include <unistd.h>
8.  #include <string.h>
9.
10. #define QUEUE_NAME    "/my_queue"
11. #define MAX_MSG_LEN   10000
12.
13. int main(int argc, char *argv[]) {
14.     mqd_t myQ_id;
15.     char msgcontent[MAX_MSG_LEN];
16.     int msg_size;
17.     unsigned int priority;
18.
19.     /* open the queue created by the sender */
20.     myQ_id = mq_open(QUEUE_NAME, O_RDWR);
21.     if (myQ_id == (mqd_t)-1) {
22.         perror("In mq_open()");
23.         exit(1);
24.     }
25.
26.     /* retreve a message from the queue */
27.     msg_size = mq_receive(myQ_id, msgcontent, MAX_MSG_LEN,
            &priority);
28.     if (msg_size == -1) {
29.         perror("In mq_receive()");
30.         exit(1);
31.     }
32.
33.     /* output message info */
34.     printf("Received a message.\n");
35.     printf("   Content: %s\n", msgcontent);
36.     printf("   Size: %d bytes.\n", msg_size);
37.     printf("   Priority: %d\n", priority);
38.
39.     /* close the qeueu */
40.     mq_close(myQ_id);
41.
42.     return 0;
43. }
```

Figure 6.10 Receiving messages from a message.

opens the queue created by the *sender* program and then retrieves a message from the queue. It is important that both programs use the same message queue name. In addition, when the *sender* program is executed for the first time, the mq_open() routine with four arguments in Lines 18–19 is called and returns a valid queue ID. When the program is executed again, this routine will attempt

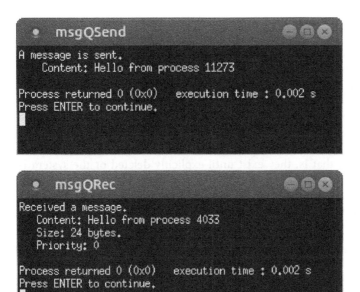

Figure 6.11 Execution of programs listed in Figures 6.9 and 6.10.

to create an existing queue and thus return −1 with the error code being EEX-IST. When this happens, we should call the mq_open () routine with two arguments to open the existing queue. That is why we have the if-statement in Lines 22–35. In the *receiver* program, we can only call the my_open() routine with two arguments.

In addition, notice that both programs call the mq_close () routine to close the queue. However, after the execution of the programs, the memory allocated in the kernel for the queue remains there. To deallocate the memory, the my_unlink () routine has to be called explicitly. We can add the call to the end of the *receiver* program or write a new program to call this routine to remove the queue from the kernel.

The execution results of the two programs are shown in Figure 6.11.

6.4.2 Shared Memory

Shared memory is a low-level way for tasks to communicate with each other. Data is exchanged by placing it in memory pages shared by multiple tasks. Shared memory is mapped into the address spaces of all tasks concerned. If one task writes a value into a particular byte of shared memory, the change is visible to other tasks immediately. This means that data transfer with shared memory is not mediated by the kernel, and thus, shared memory is a fast inter-task communication mechanism compared to message queues. This is because in message queue approach, senders need to copy messages from their local space to kernel memory, while receivers need to copy messages from kernel to their local space. Of course, additional mechanisms, such as mutexes, condition variables, and/or semaphores, have to be used to protect the access to the shared data. As with message queues, we can place any structure of data in a shared memory area.

POSIX shared memory objects are implemented as files. The objects have *kernel persistence*, that is, they exist until explicitly deleted or the system is rebooted. To set up a shared memory region, first we need to open a shared memory object, and then, we use the resulting descriptor to map the object into a task's address space. The routine used to create a new or open an existing shared memory object is

```
int shm_open(const char *name,
             int oflag,
             mode_t mode);
```

The first argument name of the routine is a reference of the shared memory object. A shared memory object name must follow the construction rules as for a normal file path name, which has to start with a "/" and cannot contain additional "/."

The second argument oflags controls the way in which the shared memory object is opened. It can be O_RDONLY for read-only or O_RDWR for read and write. Write-only is not an option for shared memory. The flag can be OR'ed with O_CREATE, meaning that the routine is called to create a shared memory object. Only when O_CREATE is used, the last argument mode is necessary. It may also be OR'ed with other flags. For example, you can specify O_EXCL to modify the behavior of O_CREATE: if the object referenced by name already exists and O_CREATE is set, but O_EXCL is not set in the shm_open() call, then the call will simply return the descriptor of the existing object. If both flags are set, then the call will return an error.

The last argument mode specifies access permission bits for the shared memory object.

Upon success, shm_open() returns a nonnegative file descriptor. Upon failure, shm_open() returns −1.

To map a shared memory into the address space of the calling process, we call the mmap() routine:

```
void *mmap(void *addr,
           size_t length,
           int prot,
           int flags,
           int fd,
           off_t offset);
```

It works by creating a virtual memory mapping for a region referenced by the file descriptor fd, which is the return value of the shm_open() call. The area of the shared memory from offset offset and with length length will be mapped into the virtual address space, specified by addr, of the process. The third argument prot specifies memory protection mode that must be consistent with the access mode specified in the shm_open() routine. The argument flags has only one option MAP_SHARED, which makes the caller's modifications to the mapped memory visible to other processes mapping the same object.

Other important routines are listed as follows:

- int munmap(void *addr, size_t length) It removes any mappings for those entire pages containing any part of the address space of the process starting at addr and continuing for length bytes.
- int close(int fd) It closes the file descriptor fd returned by the shm_open() call.
- int shm_unlink(const char *name) It remove the object name, marks it for deletion once all processes have closed it.
- fstat(int fd, struct stat *buf) It obtains information about an open file associated with the file descriptor fd and writes it to the area pointed by buf.

Note that struct stat is a system struct that is defined to store information about files. It is used in several system calls, such as fstat, lstat, and stat.

Figure 6.12 lists a program (*writer*) that creates a shared memory region (Line 19), configures its size (Line 26), maps the shared memory to its local space (Line 29), and then writes data to the shared memory (Line 40). Figure 6.13 lists a program (*reader*) that opens the share memory created by the *writer* program (Line 16), maps the shared memory to its local space (Line 23), reads data from the shared memory (Line 31), and then removes the shared memory segment (Line 34). Figure 6.14 shows the execution results of the two programs.

6.4.3 Shared Memory Protection

Processes that access shared memory need to be synchronized so that each process does not step on another process's work in the shared memory. The mutex, condition variable, and semaphore approaches introduced earlier in this chapter are only suitable for synchronization between threads of a single process. One way to synchronize data sharing between processes is to use *named semaphores*.

Named semaphores obey all the name rules as message queues. To open an existing semaphore, we use the sem_open function, with the semaphore's name and normal flags as arguments:

```
sem_t *sem_open(const char *name, int oflag);
```

```
1.  #include <stdio.h>
2.  #include <stdlib.h>
3.  #include <unistd.h>
4.  #include <sys/types.h>
5.  #include <fcntl.h>
6.  #include <sys/shm.h>
7.  #include <sys/mman.h>
8.
9.  int main(){
10.     const int SHM_SIZE = 4096;
11.     const int MSG_SIZE = 100;
12.     const char *name = "/my_shm";
13.     char message[MSG_SIZE];
14.
15.     int shm_fd;
16.     void *ptr;
17.
18.     /* create the shared memory segment */
19.     shm_fd = shm_open(name, O_CREAT | O_RDWR, 0666);
20.     if (shm_fd == -1){
21.         perror("In shm_open()");
22.         exit(1);
23.     }
24.
25.     /* configure the size of the shared memory segment */
26.     ftruncate(shm_fd,SHM_SIZE);
27.
28.     /* now map the shared memory segment in the address space
            of the process */
29.     ptr = mmap(0,SHM_SIZE, PROT_READ | PROT_WRITE, MAP_SHARED,
            shm_fd, 0);
30.     if (ptr == MAP_FAILED) {
31.         printf("Map failed\n");
32.         return -1;
33.     }
34.
35.     /* input a message from keyboard */
36.     printf("Type a message:\n");
37.     fgets (message, MSG_SIZE, stdin);
38.
39.     /* write the message to the shared memory region */
40.     sprintf(ptr,"%s",message);
41.
42.     printf("Your message has been written to the shared
            memory.\n");
43.     printf("   Content: %s\n", message);
44.     return 0;
45. }
```

Figure 6.12 Writing to a shared memory region.

```
1.  #include <stdio.h>
2.  #include <stdlib.h>
3.  #include <fcntl.h>
4.  #include <sys/shm.h>
5.  #include <sys/mman.h>
6.
7.  int main()
8.  {
9.      const char *name = "/my_shm";
10.     const int SIZE = 4096;
11.
12.     int shm_fd;
13.     void *ptr;
14.
15.     /* open the shared memory segment */
16.     shm_fd = shm_open(name, O_RDONLY, 0666);
17.     if (shm_fd == -1) {
18.         perror("in shm_open()");
19.         exit(1);
20.     }
21.
22.     /* now map the shared memory segment in the
           address space of the process */
23.     ptr = mmap(0,SIZE, PROT_READ, MAP_SHARED, shm_fd, 0);
24.     if (ptr == MAP_FAILED) {
25.         perror("in mmap()");
26.         exit(1);
27.     }
28.
29.     /* now read from the shared memory region */
30.     printf("Content in the shared memory:\n");
31.     printf("    %s", ptr);
32.
33.     /* remove the shared memory segment */
34.     if (shm_unlink(name) == -1) {
35.         perror("in shm_unlink()");
36.         exit(1);
37.     }
38.
39.     return 0;
40. }
```

Figure 6.13 Reading from a shared memory region.

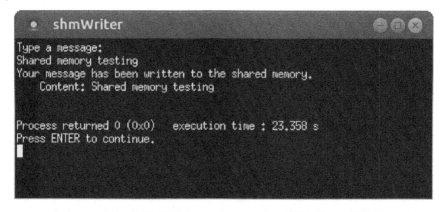

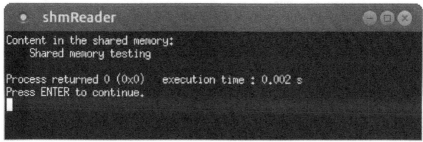

Figure 6.14 Execution of programs listed in Figures 6.12 and 6.13.

To create a new named semaphore, the same function is used, but with additional two arguments:

```
sem_t *sem_open(const char *name, int oflag,
              mode_t mode, unsigned int value);
```

In the two types of sem_open function calls, there are only two flags that we can set, and both are related to semaphore creation: O_CREAT and O_EXCL. When they are set, the call with four arguments should be used.

Other two functions related to named semaphores are sem_close and sem_unlink. sem_close removes a process's connection to the specified semaphore, while sem_unlink destroys a semaphore. The behaviors of sem_wait and sem_post are exactly the same as those used for unnamed semaphores.

The programs listed in Figures 6.12 and 6.13 are not safe, because access to the shared memory is not synchronized. Figure 6.15 lists the code of writing to shared memory with semaphore control, in which semaphore-related code is highlighted with boldface. The reading program listed in Figure 6.13 can be modified identically.

```
1.  #include <stdio.h>
2.  #include <stdlib.h>
3.  #include <unistd.h>
4.  #include <sys/types.h>
5.  #include <fcntl.h>
6.  #include <sys/shm.h>
7.  #include <sys/mman.h>
8.  #include <semaphore.h>
9.
10. int main(){
11.     const int SHM_SIZE = 4096;
12.     const int MSG_SIZE = 100;
13.     const char *shm_name = "/my_shm";
14.     const char *sem_name = "/my_sem";    /* semaphore name */
15. char message[MSG_SIZE];
16.
17.     int shm_fd;
18.     sem_t *sem;   //semaphore descriptor
19.     void *ptr;
20.
21.     /* create the shared memory segment */
22.     shm_fd = shm_open(shm_name, O_CREAT | O_RDWR, 0666);
23.     if (shm_fd == -1){
24.         perror("In shm_open()");
25.         exit(1);
26.     }
27.
28.     /* create the named semahphore. initial value: 1 */
29.     sem = sem_open(sem_name, O_CREAT, 0664, 1);
30.
31.     /* configure the size of the shared memory segment */
32.     ftruncate(shm_fd,SHM_SIZE);
33.
34.     /* now map the shared memory segment in the address space
              of the process */
35.     ptr = mmap(0,SHM_SIZE, PROT_READ | PROT_WRITE, MAP_SHARED,
              shm_fd, 0);
36.     if (ptr == MAP_FAILED) {
37.         printf("Map failed\n");
38.         return -1;
39.     }
40.
41. /* input a message from keyboard */
42.     printf("Type a message:\n");
43.     fgets (message, MSG_SIZE, stdin);
44.
45. /* write the message to the shared memory region */
46. /* access controlled by the named semaphore */
```

Figure 6.15 Writing to a shared memory region with semaphore control.

```
47.    sem_wait(sem);
48.        sprintf(ptr,"%s",message);
49.    sem_post(sem);
50.
51.    printf("Your message has been written to the shared
          memory.\n");
52.    printf("   Content: %s\n", message);
53.
54.    sem_close(sem);  /* close the semaphore */
55.    sem_unlink(sem_name);  /* destroy the semaphore */
56. return 0;
57. }
```

Figure 6.15 *(Continued)*

6.5 Real-Time Facilities

In a real-time system, the majority of tasks are periodic tasks. To implement these tasks, we need to have an effective way of tracking the passage of time. This is also important to make sure that no tasks will overrun. In this section, we introduce widely used real-time signals and timers. We also discuss how to use these kernel facilities to implement periodic tasks.

6.5.1 Real-Time Signals

Similarly to mutexes, conditional variables, semaphores, message queues, and shared memory, signals are an integral part of multitasking of many real-time kernels. Signals can be used for several different purposes, such as exception handling, process termination in abnormal circumstances, and even intertask communication. In this section, we focus on process notification of asynchronous event occurrence, particularly timer expiration.

A POSIX signal is the software equivalent of an interrupt. A signal is an asynchronous notification sent to a process or to a specific thread within the same process in order to notify it of the occurrence of an event. Here, asynchronous means that the event can occur at any time that may be unrelated to the execution of the process. An example is the key stroke of CTRL+C.

Signals can be originated by the kernel, terminal driver, or other processes. For example, the Unix command

```
$ kill -KILL 1234
```

sends a SIGKILL signal to a process whose ID is 1234. Signals are identified by their numbers. Each POSIX compliant system supports a list of signal numbers. Normally, the head file `signal.h` defines symbolic names for signals. Each signal number has a particular meaning and effect on the process that receives the signal. For example, if you hit CTRL+C, a signal of SIGINT from the OS is generated. When a process makes illegal memory reference, the event gains attention of the OS, and the OS stops the application process immediately and sends a SIGSEGV signal out. The signal will be caught by the default signal handler of SIGSEGV, which prints an error message and exits the process.

User processes can send signals to other processes through `kill()` or `sigqueue()` call. Upon receiving a signal, a process can ignore the signal, block the signal, or handle the signal. Some signals, such as `SIGKILL` (terminating a process) and `SIGSTOP` (pausing a process), cannot be ignored though.

When real-time signals are generated as a result of a POSIX timer, upon completion of asynchronous I/O, or by arrival of a message on an empty message queue, there is no server process to send the signals. Instead, the way that signals are to be delivered is set as part of initialization of the timer, the asynchronous I/O, or the message queue, by using the data structure `sigevent`.'

```
union sigval {   /* data passed with notification */
        int      sival_int;   /* integer value */
        void    *sival_ptr;   /* points to timer_id */
};

struct sigevent {
    int sigev_notify; /* notification method. */
    int sigev_signo;  /* notification signal */
    union sigval sigev_value;
                /* data to pass with notification */
};
```

The `sigev_notify` field in `struct sigevent` specifies how notification is to be performed. When it is set to `SIGEV_NONE`, then no signal is sent when the event occurs; if it is set to `SIGEV_SIGNAL`, then the process is notified to send the signal that is specified in the `sigev_signo` argument. In case of timer expiration, the signal should be `SIGALRM`. The third argument `sigev_value` is an application-defined value to be passed to a particular signal handler at the time of signal delivery. In case of timer expiration, we only need to set the first two members in structure.

To suspend a process until one of the expected signals is pending, we can call the `sigwait()` routine:

```
int sigwait(const sigset_t *set, int *sig);
```

The first argument of the routine is a signal set. The second argument stores the signal that is received. The routine returns 0 if the call is successful. Otherwise, it returns a positive error number.

6.5.1.1 Blocking Signals

To *block* a signal is to *queue* it for delivery at a later time. The purpose of blocking signals is to avoid race conditions – a signal of type X arrives while the handler for signals of type X is executing. When the signal handler returns, the block is removed for the signal in front of the queue. The POSIX signal system uses *signal sets* to deal with pending signals that might otherwise be missed while a signal is being processed. It provides several functions for creating, changing, and examining signal sets, and they are all included in `signal.h`.

- `int sigemptyset(sigset_t *set)` Initialize a signal set to be empty.
- `int sigfillset(sigset_t *set)` Initialize a signal set to be full.
- `int sigaddset(sigset_t *set, int signo)` Add the signal numbered `signo` to the specified set.
- `int sigdelset(sigset_t *set, int signo)` Remove the signal numbered signo from the specified set.
- `int sigismember(const sigset_t *set, int signo)` Check whether the signal numbered `signo` is in the specified set.

The collection of signals that are currently blocked is called the *signal mask*. Each process has its own signal mask in the kernel. When a new process is created, it inherits its parent's mask. Signals can be blocked or unblocked by modifying the signal mask. Signal mask is manipulated and interrogated by the `sigprocmask()` routine:

```
int sigprocmask(int iHow, const sigset_t *psSet,
                sigset_t *psOldSet);
```

In this routine, the argument `psSet` points to a signal set. The first argument modifies the signal mask. It can be set to one of the following three values:

- `SIG_BLOCK`: Add psSet to the current mask.
- `SIG_UNBLOCK`: Remove psSet from the current mask.
- `SIG_SETMASK`: Install psSet as the signal mask.

The last argument, `psOldSet`, is used to store the old process signal mask.
For example, the following code segment blocks signal SIGINT.

```
...
sigset_t sSet;
sigemptyset(&sSet);
sigaddset(&sSet, SIGINT);
sigprocmask(SIG_BLOCK, &sSet, NULL);
    ...
```

Note that `sigprocmask()` is only used for single-threaded process. In multithreaded processes, `pthread_sigmask()` should be used.

6.5.1.2 Dealing with Signals

Signals can be handled with default actions or user-defined handlers. To handle signals of certain type with a user-defined handler, a function needs to be set up so that it is called whenever a signal with a particular number arrives. The details of what the process needs to do upon receiving a signal are set in the sigaction structure:‘

```
struct sigaction {
    void (*sa_handler)(int); /* address of signal handler */
    sigset_t sa_mask; /* signals */
    int sa_flags;    /* signal options */
    void (*sa_sigaction)(int, siginfo_t *, void*);
                           /* alternate signal handler */
};
```

The pointer sa_handler points to a function that serves as the signal handler. The only argument (integer) of the signal handler is the signal number. The pointer sa_sigaction points to a function that serves as an alternative signal handler. Normally, we do not assign both sa_handler and sa_sigaction.

To change the action taken by a process upon receipt of a specific signal, we can call the sigaction routine:

```
int sigaction(int signum, const struct sigaction *act,
              struct sigaction *oldact);
```

Here, signum specifies the signal and can be any valid signal except SIGKILL and SIGSTOP. If act is non-NULL, the new action for signal signum is installed from act. If oldact is non-NULL, the previous action is saved in oldact.

Figure 6.16 lists a simple program that captures the CRTL+C (to terminate a process) and CRTL-Z (to suspend a process) key strokes and sends a signal to the signal handler my_handler, which outputs a message.

As restricted by the sigaction structure definition and specified in Line 5, my_handler has only one integral argument signo, which is the signal number. In the main function, we first declared an instance of the sigaction structure, named action. Data members of action are set in Lines 19–29. Lines 28 and 29 specify that the signals generated from CRTL-C and CRTL-Z key strokes will be handled by the sigaction instance action. The while-loop simply lets the process run forever so that users can type and observe the results. Figure 6.17 shows a screenshot of the execution result of the program.

6.5.2 Timers

To time a process's execution so that it runs at certain frequency, real-time clocks and timers are necessary.

All Unix-like systems use Unix time, which is also known as POSIX time or Epoch time. It is a system for describing instants in time as the number of

```
1.  #include <stdio.h>
2.  #include <signal.h>
3.  #include <unistd.h>
4.
5.  void my_handler(int signo){
6.      /* handling Ctrl-C */
7.      if (signo == SIGINT)
8.          printf("You hit Ctrl-C. \n");
9.
10.     /* handling Ctrl-Z */
11.     if (signo == SIGTSTP)
12.         printf("You hit Ctrl-Z. \n");
13. }
14.
15. int main(void){
16.     struct sigaction action;
17.
18.     /* set up signal handler */
19.     action.sa_handler = my_handler;
20.
21.     /* initialize signal set */
22.     sigemptyset(&action.sa_mask);
23.
24.     /* set signal option to 0 that makes no change to
           signal behavior */
25.     action.sa_flags = 0;
26.
27.     /* specify signals to be handled by action */
28.     sigaction(SIGINT, &action, NULL);
29.     sigaction(SIGTSTP, &action, NULL);
30.
31.     /* wait forever */
32.     while(1)
33.         sleep(1);
34.
35.     return 0;
36. }
```

Figure 6.16 Handling signals generated from key strokes of CTRL-C and CTRL-Z.

Figure 6.17 A screenshot of executing the program listed in Figure 6.16.

seconds that have elapsed since 00:00 AM, January 1, 1970. The call of the time() routine with a NULL argument returns the current time:

```
#include <time.h>
time_t time(time_t *what_time_it_is);
```

If a pointer is passed to the call, the returned time will be stored in the memory referenced by the pointer. The POSIX routine clock_gettime() can return the time with a precision that is measured in nanoseconds:

```
int clock_gettime(clockid_t c_id,
                   struct timespec *current_time);
```

After the call, the current clock time is stored in an object of timespec referenced by current_time. The timespec structure is defined as follows:

```
struct timespec {
     time_t tv_sec;    /* seconds */
     time_t tv_nsec;   /* nanoeconds */
};
```

We can also call the clock_getres() routine to obtain the clock resolution.

Real-time applications often schedule actions using *interval timers*. An interval timer can be either of two types: *one-shot* or *periodic*. A one-shot timer is an armed timer that is set to an expiration time relative to either the current time or an absolute time. When it expires, the timer is disarmed. Such a timer is useful for single-shot tasks, such as clearing buffers after the data has been transferred to storage or to time-out an operation. A periodic timer is armed with an initial expiration time, absolute or relative, and a repetition interval. Each time the interval timer expires, it is reloaded with the interval and rearmed. This timer is useful for periodic tasks.

POSIX defined a set of routines for timers that utilize Unix clock. The most fundamental one is

```
int timer_create(clock_id clockid, struct sigevent sigev,
                 timer_t *timerid);
```

which is called to create a new per-process interval timer. The clockid argument specifies the clock that the new timer uses to measure time. All POSIX-compliant RTOSs must support CLOCK_REALTIME, which is a settable system-wide real-time clock. When the timer is successfully created, the ID of the new timer is returned in the buffer pointed by timerid, which must be a non-NULL pointer. This ID is unique within the process, until the timer is deleted. The new timer is initially disarmed.

The second argument `sigev` is a pointer to the `struct sigevent` data structure. This data structure is used to inform the kernel about what kind of event the timer should deliver whenever it "fires." In our case, we set the first two members of the structure as follows:

```
sigev.sigev_notify = SIGEV_SIGNAL;
sigev.sigev_signo = SIGALRM;
```

After a timer is created, we need to set up the timer. The routine is

```
int timer_settime(timer_t timerid, int flags,
          const struct itimerspec *new_setting,
          struct itimerspec *old_setting);
```

It sets up the timer referenced by `timerid` to expire either periodically or once. The last two arguments are pointers of `itimerspec` structure, which is defined as follows:

```
struct itimerspec {
    struct timespec it_interval;  /* Timer interval *
    struct timespec it_value;     /* Initial expira-
tion */
};
```

To set a timer, we set `new_setting->it_value` with the time interval after which the timer should expire for the first time and set `new_setting->it_interval` with the interval at which subsequent timer expirations should occur. If `new_setting->it_value` is set to 0, then the timer will never expire; if `new_setting->it_interval` is set to 0, then the timer will expire only once, at the time indicated by `it_value`.

If the *flags* argument is set to 0, then the `new_setting->it_value` field is taken to be a time relative to the current time. If it is set to `TIMER_ABSTIME`, then the time is absolute.

Other timer-related POSIX routines include the following:

- `int timer_delete(timer_t timerid)` Deletes the timer whose ID is given in `timerid`.
- `int timer_getoverrun(timer_t timerid)` Returns the overrun count of the timer whose ID is given in `timerid`.

Note that in calls to interval timer functions, time values smaller than the resolution of the system hardware periodic timer are rounded up to the next multiple of the hardware timer interval. For example, if the clock resolution is 10 milliseconds and the value set for timer expiration is 95 milliseconds, then the timer will expire in 100 milliseconds, instead of 95 milliseconds.

6.5.3 Implement Periodic Tasks

When we implement a periodic task, we need to make sure that the task starts to execute repeatedly at the beginning of each period, and it is suspended each cycle from the time point when it is completed to the beginning of the next cycle. The framework of periodic task implementation is as follows:

```
aPeriodicTask{
    initialize phase, period, etc.;
    set_timer(phase, period);
    while (condition) {
        task_body();
        wait_next_activation();
    }
}
```

In the aforementioned framework, the real-time control of the task is carried out through two actions: one is to set a timer to wake up at a given period, the function call of set_timer(phase, period). The second action is to put the task to wait until the next period begins, the function call of wait_next_activation(). We discuss how to implement these two actions in the rest of this section. Typically, the task body performs routine jobs listed as follows:

```
task_body(){
    receive data;
    computation;
    update state variables;
    output data;
}
```

The actual implementation varies with the mission of each individual task.

6.5.3.1 Using sleep() Function

A simple idea to make a task run periodically is to call the sleep() or similar routines after a task instance is completed to make the task sleep until the next period begins. With this approach, the set-timer action is not needed. The wait-next action is implemented as follows:

```
wait_next_activation(){
    current_time = time();
    sleep_time = next_activation_time - current_time;
    next_activation_time = next_activation_time + period;
    sleep(sleep_time);
}
```

```
1.  #include <sys/time.h>
2.  #include <signal.h>
3.  #include <time.h>
4.  #include <stdlib.h>
5.  #include <stdint.h>
6.  #include <string.h>
7.  #include <stdio.h>
8.
9.  #define ONE_THOUSAND 1000
10. #define ONE_MILLION 1000000
11. /* offset and period are in microseconds. */
12. #define OFFSET 1000000
13. #define PERIOD 500000
14.
15. sigset_t sigst;
16.
17. static void wait_next_activation(void){
18.     int dummy;
19.     /* suspend calling process until a signal is pending */
20.     sigwait(&sigst, &dummy);
21. }
22.
23. int start_periodic_timer(uint64_t offset, int period){
24.     struct itimerspec timer_spec;
25.     struct sigevent sigev;
26.     timer_t timer;
27.     const int signal = SIGALRM;
28.     int res;
29.
30.     /* set timer parameters */
31.     timer_spec.it_value.tv_sec = offset / ONE_MILLION;
32.     timer_spec.it_value.tv_nsec = (offset % ONE_MILLION) *
            ONE_THOUSAND;
33.     timer_spec.it_interval.tv_sec = period / ONE_MILLION;
34.     timer_spec.it_interval.tv_nsec = (period % ONE_MILLION) *
            ONE_THOUSAND;
35.
36.     sigemptyset(&sigst);   /* initialize a signal set */
37.     sigaddset(&sigst, signal);   /* add SIGALRM to the
            signal set */
38.     sigprocmask(SIG_BLOCK, &sigst, NULL);   /* block the signal */
39.
40.     /* set the signal event at timer expiration */
41.     memset(&sigev, 0, sizeof(struct sigevent));
42.     sigev.sigev_notify = SIGEV_SIGNAL;
43.     sigev.sigev_signo = signal;
44.
45.     /* create timer */
46.     res = timer_create(CLOCK_MONOTONIC, &sigev, &timer);
```

Figure 6.18 Implementing a periodic task.

```
47.
48.      if (res < 0) {
49.          perror("Timer Create");
50.          exit(-1);
51.      }
52.
53.      /* activiate the timer */
54.      return timer_settime(timer, 0, &timer_spec, NULL);
55. }
56.
57. static void task_body(void){
58.      static int cycles = 0;
59.      static uint64_t start;
60.      uint64_t current;
61.      struct timespec tv;
62.
63.      if (start == 0) {
64.          clock_gettime(CLOCK_MONOTONIC, &tv);
65.          start = tv.tv_sec * ONE_THOUSAND + tv.tv_nsec / ONE_MILLION;
66.      }
67.
68.      clock_gettime(CLOCK_MONOTONIC, &tv);
69.      current = tv.tv_sec * ONE_THOUSAND + tv.tv_nsec / ONE_MILLION;
70.
71.      if (cycles > 0){
72.          printf("Ave interval between instances: %f milliseconds\n",
73.                              (double)(current - start)/cycles);
74.      }
75.
76.      cycles ++;
77. }
78.
79. int main(int argc, char *argv[]){
80.      int res;
81.
82.      /* set and activate a timer */
83.      res = start_periodic_timer(OFFSET, PERIOD);
84.      if (res < 0) {
85.          perror("Start Periodic Timer");
86.          return -1;
87.      }
88.
89.      while(1) {
90.          wait_next_activation(); /* wait for timer expiration */
91.          task_body();  /* executes the task */
92.      }
93.
94.      return 0;
95. }
```

Figure 6.18 (*Continued*)

This solution is not reliable. If the process or thread is preempted after the sleep time is calculated and before the sleep() call is performed, then the process will "over sleep." For example, if the calculated sleep time is 5 milliseconds and the time-span from the process being preempted to getting resumed is 3 milliseconds, then the process should only sleep for 2 milliseconds, instead of 5 milliseconds.

6.5.3.2 Using Timers

Another way to implement periodic tasks is using real-time timers. In this approach, we create a timer and arm the timer in the set-timer action, and wait for the timer expiration signal in the wait-next action. The pseudocode of set-timer is as follows:

```
set_timer(phase, period){
        set up an itimerspec instance with phase and period;
        add SIGALRM to an empty signal set and mask it;
        set up a sigevent instance with signal SIGALRM
            and notification method SIGEV_SIGNAL;
        create a timer with the sigevent instance;
        arm the timer with phase and period;
}
wait_next_activation(){
        wait for signal SIGALRM;
}
```

Figure 6.18 lists a program that implements the pseudocode listed earlier. The timer is set to expire every 500 milliseconds. The offset is set to 1000 milliseconds, meaning that the first expiration of the timer occurs 1500 milliseconds after the timer is activated. In the main function, we first call the function `start_periodic_timer()` to set and activate the timer. Then the process enters an infinite loop, in which the function `wait_next_activation()` is called to wait for the next timer expiration signal first. When the signal arrives, the call returns, and the function `task_body()` is called and executed, which simply calculates and prints out the average interval between consecutive instances since the first expiration of the timer. Figure 6.19 is a screenshot of the output of the program execution.

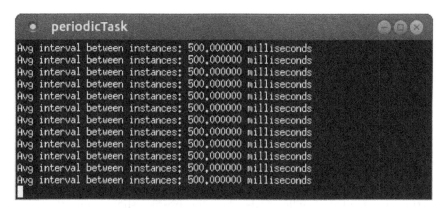

Figure 6.19 A screenshot of the output of the program listed in Figure 6.17.

6.5.4 Implement an Application with Multiple Periodic Tasks

In the real world, even a simple real-time application would have more than one periodic task. A periodic task can be implemented with a process or a thread, depending on the computation complexity of the task. Simple tasks are usually implemented with threads. Regardless, each periodic task would need a timer created and activated. Each timer takes the corresponding task's period and phase as its parameters. Multiple timers can be created in a program (process), each for a task (thread). The common definitions and utilities for timer setup and activation and other application related functions should be placed in a header file, while each thread should contain only the code that is unique to the corresponding task.

Exercises

1 Write a simple C program using Pthread that creates a thread with a default attribute object. All the thread does is create an array of 1,000,000 elements of integer type, initialize the array with random numbers between 0 and 100, and then output a message indicating that the initialization is completed. Run the program and see what happens. Then, modify your code such that you create the thread with a customized attribute object, in which you specify the stack size as 20 MB. Run the program again and see what happens. Explain the difference.

2 When creating a new Pthread by calling the `pthread_create()` routine, we can pass one and only one argument to the thread start routine. The argument must be cast to `(void *)`. The following call passes a single integer t to the thread start routine myRoutine:

```
rc = pthread_create(&thread, NULL, myRoutine, (void *)t);
```

To retrieve t in myRoutine, we perform a cast as follows:

```
void *myRoutine(void *arg)
{
    int a = (int) arg;   //a gets the value of t
    . . .
}
```

The following code attempts to pass the address of an integer to the start routine. Is it safe?

```
for(t=0; t<NUM_THREADS; t++)
{
```

```
printf("Creating thread %ld\n", t);
rc = pthread_create(&th[t], NULL, myRoutine,
(void *) &t);
    . . .
}
```

3 If multiple data items are to be passed to a thread during the thread creation, we can define a structure that contains all of the data items as members, declare an instance with desired values for all members, and then pass the reference of the instance to the thread start routine. Write a program that passes the following data to a thread:

```
int      student_id;
char    *first_name;
char    *last_name;
double   gpa;
```

Initialize the data in the main function and output the data in the thread start routine.

4 The program listed in Figure 6.19 creates three Pthreads, each thread containing a for-loop. A global variable count is used to track the total iteration times of all the three threads. For each loop, a thread reads the global count into a local count, increments the local count, and then writes it to the global count. Run the program multiple times, and check to see if the value of the global count is always correct. If it is not, explain why, and then fix the problem.

```
1.  #include <pthread.h>
2.  #include <stdio.h>
3.  #include <stdlib.h>
4.
5.  #define ITERATION 100000
6.
7.  int count = 0;
8.
9.  void * Count(void * a){
10.     int i, local_count;
11.     for(i = 0; i < ITERATION; i++){
12.       local_count = count;
                /* copy the global count to a local_count */
13.       local_count++;
                /* increment the local_count */
14.       count = local_count;
                /* store the local value into the global count */
15.     }
16. }
```

```
17.
18. int main(int argc, char * argv[]){
19.     pthread_t tid[3];
20.     int i, correctCount;
21.
22.     for (i = 0; i <3; i++){
23.         if(pthread_create(&tid[i], NULL, Count, NULL)){
24.             printf("\n ERROR creating thread 1");
25.             exit(1);
26.         }
27.     }
28.
29.     for (i = 0; i <3; i++){
30.         if(pthread_join(tid[i], NULL)){
31.             printf("\n ERROR joining thread");
32.             exit(1);
33.         }
34.     }
35.
36.     correctCount = 3 * ITERATION;
37.     if (count < correctCount)
38. printf("\n BOOM! count is %d, should be %d\n",
        count, correctCount);
39.     else
40. printf("\n OK! cnt is %d\n", count);
41.
42.     pthread_exit(NULL);
43. }
```

5 Assume that we are given an array of 200,000 elements of double type, and we want to find the sum of all elements with two threads. The first thread finds the sum of the first half of the array and adds it to a global variable total. Similarly, the second thread finds the sum of the second half of the array and adds it to total. Write a program to implement the function. Declare the array as global data, and initialize it with all 0.99s in the main function before threads are created. The access to total should be protected by a mutex object.

6 Perform the previous project, but generalize the solution by specifying the number of threads to be created and dividing the work equally to each thread. Make sure that the number of threads can divide the array size.

7 Write a program that creates a single Pthread. Implement the "join" function using a condition variable and mutex, such that the main thread will wait until the child Pthread is completed before it exits, instead of calling the pthread_join routine.

8 Redo the previous problem with a semaphore, instead of a condition variable and mutex.

9 Write a program that uses two threads to search a doubly linked list. Requirements are as follows:

(1) Create a doubly linked list. Insert 10,000 nodes to the list. Each node stores a randomly generated integral value in the range of 0...50,000.

(2) Create two Pthreads. One searches the list for a node that stores a value x from the head toward the tail, while the other searches the list for the node backward from the tail. When one thread hits the node, both threads should stop searching.

(3) When the search is stopped, each thread prints out the number of nodes it has examined.

(4) The value x to be searched for is entered by the user. It must be in the range of 0...50,000.

(5) If there are multiple nodes that store x in the list, the search stops when the first node is hit. If there is no single node that stores x, the search stops when the entire list is examined by a thread.

10 Modify the program listed in Figure 6.5 such that the three Pthreads keep running until the main thread explicitly calls the `pthread_cancel()` routine to terminate them 5 seconds after they are created.

11 If the semaphore is initialized to 2, what will be the semaphore value and consequence after each action listed in Table 6.2? Redraw the table.

12 Rewrite the code listed in Figure 6.3 by solving the synchronization issue with a semaphore, instead of a mutex object.

13 Modify the programs listed in Figures 6.9 and 6.10, such that the sender repeatedly takes messages (strings) from the keyboard and sends it to the message queue, until a "stop" is taken and sent out. Meanwhile, the receiver keeps receiving messages from the queue, until a "stop" message is received.

14 On top of the previous project, add two more senders, such that three senders send messages to one receiver via one message queue. Assign the highest priority to messages from the first sender, a medium priority to the messages from the second sender, and the lowest priority to the messages from the third sender. Run the senders first until all messages are sent out in a mixed order. Then, run the receiver and check to see if all messages are received in the order of their priorities.

15 Develop a project with two periodic tasks. Task 1 opens a shared memory to store an array of 1000 elements of integer type and initialize all elements to 0s. It inserts a new value that is randomly generated in the range of 1–100 to the array every 500 milliseconds. Task 2, on the other hand, counts the number of values inserted in the array every 400 milliseconds and displays the count.

16 Develop a project with three periodic tasks, where one task plays the role of a server and the other two tasks are clients. The server task opens two message queues, one for each client. Each client sends the POSIX time in a message to the server every 500 milliseconds. Meanwhile, the server retrieves a message from each message queue and displays them every 500 milliseconds.

Suggestions for Reading

The requirement of mutual exclusion was first identified and solved by Edsger W. Dijkstra [1], which is credited as the first topic in the study of concurrent algorithms. POSIX programming was discussed in details in Refs [2–4]. Bruno [5] and Wellings [6] introduced concurrent and real-time programming in Java. Burns [7] introduced concurrent and real-time programming in Ada.

References

1 Dijkstra, E.W. (1965) Solution of a problem in concurrent programming control. *Communications of the ACM*, **8** (9), 569.
2 Buttlar, D. (1996) PThreads Programming: A POSIX Standard for Better Multiprocessing *(A Nutshell Handbook)*, O'Reilly Media.
3 Butenhof, D.R. (1997) *Programming with POSIX Threads*, Addison-Wesley Professional.
4 Gallmeister, B. (1995) *POSIX.4: Programming for Real World*, O'Reilly and Associations, Inc..
5 Bruno, E.J. (2009) *Real-Time Java Programming: With Java RTS*, Prentice Hall.
6 Wellings, A.J. (2004) *Concurrent and Real-Time Programming in Java*, Wiley.
7 Burns, A. (2007) *Concurrent and Real-Time Programming in Ada*, 3rd edn, Cambridge University Press.

7

Finite-State Machines

Real-time embedded systems are reactive systems. Their primary purpose is to respond to or react to signals from their environment. The design of these systems is a complex process, which necessitates the integration of common design methods in both hardware and software to fulfill the functional and nonfunctional requirements. Design patterns, which give abstract solutions to commonly recurring design problems, have been widely used in the software and hardware domains. Finite-state machines (FSMs) are a powerful mathematical and graphical tool in specifying the behavior of reactive systems. Over the decades, they have become one of the most popular design patterns for real-time embedded systems. This chapter introduces traditional FSMs.

7.1 Finite State Machine Basics

An FSM is an abstract computation model that is used to design both software and sequential logic circuits. In general, a state machine is a device that stores the context of an object at a given time and operates on input events to change the context and/or cause an action to take place or an output to occur for any given change. The context is called a *state*, which captures the relevant aspects of the object's history. An FSM can be in one of a *finite* number of states. A change from one state to another is called a *transition*. A state transition occurs because of the occurrence of some *triggering events* or *conditions*. The state it is in at a given time is called the *current state*.

An FSM can be represented by a directed graph called a *state diagram*, in which each state is represented by a node (circle) and each transition is represented by an edge.

Example 7.1 *State Diagram of a Light*
Figure 7.1 shows the state diagram of a button-controlled light. It has two states: *On* and *Off*. When the light is in the *Off* state, a *Press button* event changes the

Real-Time Embedded Systems, First Edition. Jiacun Wang.
© 2017 John Wiley & Sons, Inc. Published 2017 by John Wiley & Sons, Inc.

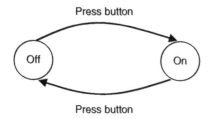

Press button

Off   On

Press button

Figure 7.1 Finite-state machine of a light.

(a)   (b)

Figure 7.2 Graphic representations of (a) initial states and (b) final states.

light state to *On*; Similarly, when the light is in the *On* state, a *Press button* event changes the light state to *Off*.

In many FSMs, it is necessary to classify states into initial states, final states, and intermediate states. Normally, an initial state is graphically marked with an incoming arrow, while a final state is represented with two circles, as shown in Figure 7.2.

Example 7.2   *State Diagram of a Safe*
Consider a safe that is locked with a code 2-0-1-7. The states that we are interested in are as follows:

- State q_0: locked, with no meaningful input.
- State q_1: locked, with an input sequence that ends with a "2."
- State q_2: locked, with an input sequence that ends with a "2-0."
- State q_3: locked, with an input sequence that ends with a "2-0-1."
- State q_4: unlocked (after an input sequence that ends with a "2-0-1-7."

Figure 7.3 shows the state diagram of the safe.

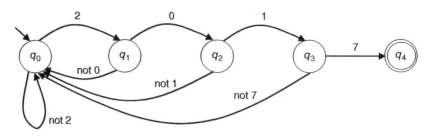

Figure 7.3 Finite-state machine of a safe.

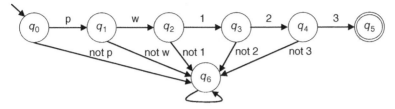

Figure 7.4 An acceptor for password pw123.

7.2 Deterministic Finite Automation (DFA)

There are two types of FSMs: *deterministic finite automation* (DFA) and *non-deterministic finite automation* (NDFA). In DFA, one can determine the state to which the machine will move for each triggering event.

A DFA is also known as a *deterministic finite accepter*, a machine that accepts and rejects finite sequence of inputs and only produces a unique computation (or run) of the automaton for each input sequence. For example, the DFA shown in Figure 7.4 will only accept a password of "pw123." All other input strings will lead the machine to the error and dead state q_6.

Mathematically, a DFA is denoted as $M = (Q, \sum, \delta, q_0, F)$, where

- Q is a finite set of states.
- $\sum$ is a finite set of symbols.
- $\delta: Q \times \sum \to Q$ is the transition function.
- q_0 is the initial state. $q_0 \in Q$.
- F is a set of final states. $F \subseteq Q$.

In the definition, Q is simply a set of a finite number of elements, with each element being the name of a state. $\sum$ is a set of finite elements as well, with each element being an input. The transition function δ is also called a *next state function*, meaning the machine moves to the state $\delta(q, \varepsilon)$ if it receives the input ε in the state q.

Example 7.3 *Mathematical Model of FSM in Figure 7.2*
The FSM shown in Figure 7.3 is a DFA. Its mathematical representation is as follows:

$$Q = \{q_0, q_1, q_2, q_3, q_4\}.$$
$$\sum = \{0, 1, 2, 3, 4, 5, 6, 7, 8, 9\}.$$
$$
\begin{aligned}
\delta: \quad & (q_0, 2) \to q_1, (q_0, \varepsilon) \to q_0 \quad \text{for all } \varepsilon \neq 2; \\
& (q_1, 0) \to q_2, (q_1, \varepsilon) \to q_0 \quad \text{for all } \varepsilon \neq 0; \\
& (q_2, 1) \to q_3, (q_2, \varepsilon) \to q_0 \quad \text{for all } \varepsilon \neq 1; \\
& (q_3, 7) \to q_4, (q_3, \varepsilon) \to q_0 \quad \text{for all } \varepsilon \neq 7.
\end{aligned}
$$
$$F = \{s_4\}.$$

The initial state is q_0.

Table 7.1 State transitions of FSM shown in Figure 7.2.

	0	1	2	3	4	5	6	7	8	9
q_0	q_0	q_0	q_1	q_0	q_0	q_0	q_0	q_0	q_0	q_0
q_1	q_2	q_0	q_0	q_0	q_0	q_0	q_0	q_0	q_0	q_0
q_2	q_0	q_3	q_0	q_0	q_0	q_0	q_0	q_0	q_0	q_0
q_3	q_0	q_0	q_0	q_0	q_0	q_0	q_0	q_4	q_0	q_0

State transitions can also be described using a *state transition table*. For example, Table 7.1 lists all state transitions shown in Figure 7.3. The rows of the table are labeled by states, the columns are labeled by inputs, and each element shows the next state.

A state machine may have outputs corresponding to each transition. There are two types of FSMs that generate outputs:

- Moore machines
- Mealy machines

A Moore machine is a DFA whose output depends only on the current state. In contrast, a Mealy Machine is a DFA whose output depends on the current input as well as the current state. Any Moore machine can be turned into a Mealy machine, and vice versa.

7.2.1 Moore Machines

Basically, a Moore machine is just a DFA with output associated with each state. The machine writes the appropriate output as it enters each state. Mathematically, a Moore machine is denoted by $M = (Q, \Sigma, O, \delta, \lambda, q_0)$, where

- Q is a finite set of states.
- Σ is a finite set of input symbols.
- O is a finite set of output symbols.
- $\delta: Q \times \Sigma \to Q$ is the transition function.
- $\lambda: Q \to O$ is the output function.
- q_0 is the initial state. $q_0 \in Q$.

Notice that there is no final state set in the Moore machine definition. This is because a Moore machine is deemed an output producer, instead of a language recognizer (acceptor). The output function λ is a mapping from the set of states to the set of outputs, indicating that outputs depend only on states. Figure 7.5 illustrates how next states and outputs are computed in a Moore machine.

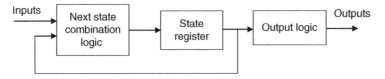

Figure 7.5 Moore machines: outputs depend solely on current states.

Example 7.4 *Moore Machine of a Vending Machine*

Consider a vending machine that sells candy bars for 20 cents each. Assume that the machine only accepts coins of 5 cents, 10 cents, and 25 cents. When coins of 20 cents or more are inserted into the machine, it releases a candy bar and ejects changes.

To model the behavior of the vending machine, we need to find out all possible states and inputs and then decide transitions among the states. All possible input events are as follows:

- 5c (insert a coin of 5 cents)
- 10c (insert a coin of 10 cents)
- 25c (insert a coin of 25 cents)

After each input, the customer's balance in the vending machine is changed. All possible balances are as follows:

- 0c (0 cents)
- 5c (5 cents)
- 10c (10 cents)
- 15c (15 cents)

In addition, when a coin is inserted, the customer may receive one of the following outputs, depending on the total number of coins inserted:

- "-" (nothing).
- "bar" (a candy bar)
- "bar, 5c" (a candy bar and a coin of 5 cents for change)
- "bar, 10c" (a candy bar and a coin of 10 cents)
- "bar, 15c" (a candy bar and coins of 15 cents)
- "bar, 20c" (a candy bar and coins of 20 cents)

However, except "-", all other outputs only come with a balance of 0c. Because outputs are only determined by states in a Moore machine, combinations of balances and outputs constitute states in this problem, which are listed in Table 7.2.

The Moore machine diagram is shown in Figure 7.6.

Consider an input sequence of 10c, 10c, 5c, and 25c, and we can easily tell from this state machine that the machine will go through states $q_0 q_7 q_3 q_6 q_3$ and sequentially output

(nothing), (nothing), a candy bar, (nothing), a candy bar plus 10c.

Table 7.2 Moore machine states of
Example 7.4.

State	Balance	Output
q_0	0c	—
q_1	0c	bar
q_2	0c	bar, 5c
q_3	0c	bar, 10c
q_4	0c	bar, 15c
q_5	0c	bar, 20c
q_6	5c	—
q_7	10c	—
q_8	15c	—

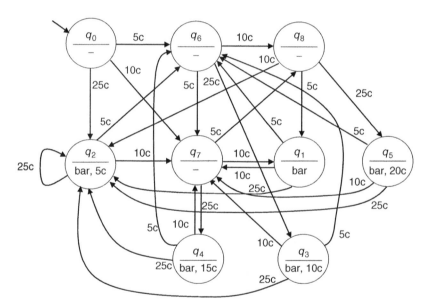

Figure 7.6 Moore machine state diagram of a vending machine.

7.2.2 Mealy Machines

Intuitively, the difference between Moore machines and Mealy machines is
that Mealy machines move the outputs from within state nodes to transitions.
Transitions in a Mealy state diagram are labeled in the format of i/o, where i is
the input and o is the output. Mathematically, a Mealy machine is denoted by
$M = (Q, \Sigma, O, \delta, \lambda, q_0)$, where

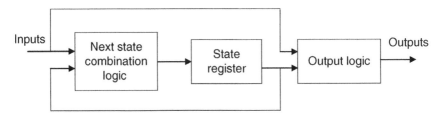

Figure 7.7 Mealy machines: outputs depend on both inputs and current states.

- Q is a finite set of states.
- $\sum$ is a finite set of input symbols.
- O is a finite set of output symbols.
- $\delta: Q \times \sum \rightarrow Q$ is the transition function.
- $\lambda: Q \times \sum \rightarrow O$ is the output function.
- q_0 is the initial state. $q_0 \in Q$.

For the same reason as with Moore machines, there are no final states defined in Mealy machines. On the other hand, the definitions of the output function λ in Mealy machines and Moore machines show the difference between the two types of machines. In Mealy machines, it is a mapping from the Cartesian product of Q and $\sum$ to O, indicating the outputs depend on both states and inputs. Figure 7.7 illustrates how next states and outputs are produced in Mealy machines.

Mealy machines are as expressive as Moore machines. However, since typically there are more transitions than states in a state machine, a Mealy machine is often more compact than a Moore machine in specifying outputs. This fact also makes Mealy machines more practical in use. For example, the behavior of the vending machine in Example 7.4 can be modeled in a Mealy machine much more concisely as shown in Figure 7.8. Basically, the six states $q_0, q_1, \ldots,$ q_5 that all have a balance of 0 cents in the Moore machine are combined into a single state marked by $0c$ in the Mealy machine model, thus reducing the total number of states.

Example 7.5 *Mealy Machine of a Seat Belt Reminder System*

A seat belt is an important vehicle safety device that prevents the occupant of a vehicle from movement in the event of a collision or a sudden stop. A seat belt reminder system gives a signal after the ignition is turned on and if the occupant's seat belt is not fastened. Assume that we want to design a seat belt reminder system that would function according to the following specification:

- The initial state is the car engine being off.
- After being seated, the driver can either turn on the engine or put the seat belt on.

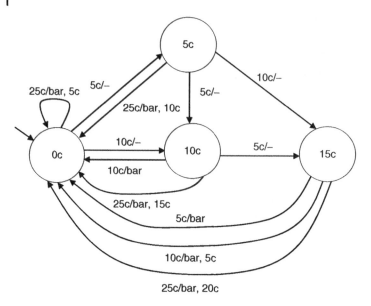

Figure 7.8 Mealy machine state diagram of the vending machine.

- When the engine is on, but the driver is seated without the seat belt buckled, the buzzer timer is turned on. The timer is turned off if the driver puts the seat belt on before the timer expires.
- If the timer expires, the buzzer is turned on. When the driver puts on his seat belt, the buzzer is turned off.
- The driver can turn off the car engine at any moment, which turns the timer or buzzer off, whichever is on.
- When the driver is seated with the belt buckled, he can take off the seat belt.
- The driver cannot turn on the car engine before he is seated.
- The driver cannot leave the seat with the seat belt buckled or while the engine is on.

The inputs of the system are from the car key, seat sensor, belt sensor, and timer. Input events are

key, seat, unseat, belt, unbelt, and *timer_expires.*

Outputs are

timer_off. timer_on, buzzer_off, and *buzzer_on.*

The system can take any one of the following states at any given time:

- *Off*: The engine is off.
- *Seated*: Driver is seated with engine off.
- *Ready*: Driver is seated with seat belt buckled and engine off.

- *Timing*: Driver is seated with engine on and timer on.
- *Belted*: Driver is belted with engine on.
- *Buzzer*: Buzzer is on.

Figure 7.9 shows the Mealy machine diagram of this seat belt system. The state transitions are also listed in Table 7.3.

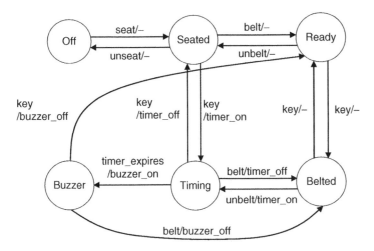

Figure 7.9 Mealy machine diagram of a seat belt system.

Table 7.3 State transition table of the Mealy machine shown in Figure 7.9.

Current State	Input	Next State	Output
Off	seat	Seated	—
Seated	unseat	Off	—
	belt	Ready	—
	key	Timing	timer_on
Ready	unbelt	Seated	—
	key	Belted	—
Belted	unbelt	Timing	timer_on
	key	Ready	—
Timing	belt	Belted	timer_off
	key	Seated	timer_off
	timer_expires	Buzzer	buzzer_on
Buzzer	belt	Belted	buzzer_off
	key	Ready	buzzer_off

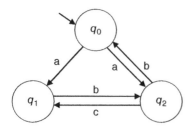

Figure 7.10 A simple NDFA.

7.3 Nondeterministic Finite Automation

In an NDFA, given a state and an input, there may be more than one next state, or a state can transition from one state to another without any input, or there is no next state at all for some given input. NDFAs are good in specifying unknown or unspecified system behavior. For any NDFA, there is always an equivalent DFA. However, the NDFA model is more compact and uses fewer states compared to the DFA model. Example 7.6 shows an NDFA.

Example 7.6   *A Simple NDFA.*
Figure 7.10 shows an NDFA, in which the input a in the state q_0 can change the system to the state q_1 or q_2. In other words, the system's behavior in response to the input event a in the initial state is nondeterministic.

7.4 Programming Finite-State Machines

As one of the most popular design patterns, FSMs are implemented in many real-time embedded applications. There are two common approaches to implementing an FSM. One is using conditional statements. An FSM can be simply coded using two levels of nested multiple-decision or switch-case structures. The first-level switch-case structure contains a list of cases corresponding to all the states. Each case within the structure would contain a second-level switch-case structure, listing the various possible inputs. Or, vice versa, one could start with the outer switch-case structure listing all the inputs, where each input case would contain a switch-case structure with a case for each state. For example, Figure 7.11 lists a code segment of the implementation in C of the FSM of the safe discussed in Example 7.2. In this simple problem, each internal switch-case structure can be replaced with an if–then statement.

The conditional statements approach is very straightforward and easy to understand. However, when the number of states and input events grow, the code can easily become unwieldy. When the state machine code runs into multiple screen pages, debugging and maintenance will become difficult, not to mention the code readability.

```
int get_input();   //get the digit that is pressed
void lock_safe();   //lock the safe
void unlock_safe();   //unlock the safe

void fsm(){
    enum states {STATE0, STATE1, STATE2, STATE3, STATE4} current_state;
    lock_safe();   //initialize the safe
    current_state = STATE0; //set the initial state
    int input;

    while(true){
     input = get_input();

     switch(current_state){
         case STATE0:
             switch(input){
                 case 2:
                     current_state = STATE1;
                     break;
                 default:
                     current_state = STATE0;
             }
         case STATE1:
             switch(input){
                 case 0:
                     current_state = STATE2;
                     break;
                 default:
                     current_state = STATE0;
             }
         case STATE2:
             switch(input){
               case 1:
                     current_state = STATE3;
                     break;
                 default:
                     current_state = STATE0;
             }
         case STATE3:
             switch(input){
               case 7:
                     current_state = STATE4;
                     break;
                 default:
                     current_state = STATE0;
             }
```

Figure 7.11 Code segment of the finite-state machine of Example 7.2.

```
        case STATE4:
            unlock_safe();
            break;
        } // switch(current_state)
    } //while(true)
}
```

Figure 7.11 (*Continued*)

Another approach to implementing state machines is table-based. To make the implementation scalable for machines with large numbers of states and input events, this approach uses a two-dimensional table, with one dimension for states and the other for events, to store transition functions. The table could be implemented in C using a two-dimensional array of function pointers.

Let us use the Mealy machine shown in Figure 7.9 as an example to explain how the table-based approach works. We define all states and input events as enumerated type:

```
enum states {OFF, SEATED, READY, BELTED, TIMING, BUZZER} current_state;
enum events {SEAT, UNSEAT, BELT, UNBELT, KEY, TIMER_EXPIRES} new_event;
```

Then, we define the state transition table as follows:

```
#define MAX_STATES 6
#define MAX_EVENTS 6
typedef void (*transition)();

transition state_table[MAX_STATES][MAX_EVENTS] = {
    {seat,  error,  error,  error,    error,    error},  // state OFF
    {error, unseat, belt_s, error,    key_s,    error},  // state SEATED
    {error, error,  error,  unbelt_r, key_r,    error},  // state READY
    {error, error,  error,  unbelt_b, key_b,    error},  // state BELTED
    {error, error,  belt_t, error,    key_t,    timer},  // state TIMING
    {error, error,  belt_b, error,    key_z,    error}}; // state BUZZER
```

Each element in the table `transition` is a function that handles the corresponding input event in the corresponding state. The `error` function is used to handle all events that are not acceptable and will be ignored in a state. For example, the state `TIMING` (`Timing` in the state machine diagram) accepts and handles three events (`BELT`, `KEY`, and `TIMER_EXPIRES`) and ignores all other inputs. Except the `error` function, each function in the table should be implemented to generate the output and next state. Here is an example of the function `belt_t()`:

```
/* Function belt_t
 * Input event: BELT       Output: timer_off
 * Current state: TIMING  Next state: BELTED */
void belt_t(){
    turn_timer_off();       //turn_timer_off() should be implemented
  somewhere
    current_state = BELTED;    //this is a global variable
}
```

The aforementioned definitions and implementation of all functions can be placed in a separated file. The body of the main program will then be simply implemented as

```
while(true) {
    new_event =  get_new_event(); /* get the next event to process */

    if ((new_event >= 0) && (new_event < MAX_EVENTS)
        && (current_state >= 0) && (current_state < MAX_STATES)) {
        /* call the transition function */
        state_table[current_state][new_event]();
    }
    else {
        /* invalid event/state - handle appropriately */
    }
}
```

Exercises

1 Create the state transition table for the Mealy machine shown in Figure 7.12a. What is the output sequence for an input sequence 0011100?

2 Create the state transition table for the Mealy machine shown in Figure 7.12b. What is the output sequence for an input sequence abaabaa?

3 Draw the Mealy machine for the state transitions shown in Table 7.4. A is the entry state of the machine.

4 Create the state transition table for the Moore machine shown in Figure 7.6.

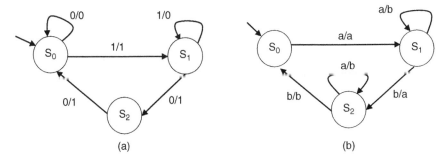

(a)   (b)

Figure 7.12 Two three-state Mealy machines.

Table 7.4 The state transition table for Problem 7.3.

Present state	Input	Next state	Output
A	a	B	b
	b	C	—
B	a	C	—
	b	B	a
C	a	A	a
	b	D	a
D	a	A	b
	b	D	—

Table 7.5 The state transition table for Problem 7.5.

Present state	Input	Next state	Output
A	a	B	—
	b	C	
B	a	C	a
	b	B	
C	a	A	a
	b	D	
D	a	A	b
	b	D	

5 Draw the Moore machine for the state transitions shown in Table 7.5. A is the entry state of the machine.

6 Draw the Moore machine for the state transitions shown in Table 7.6. A is the entry state of the machine.

7 Create the state transition table for the NDFA shown in Figure 7.10.

8 Consider a digital combination lock as illustrated in Figure 7.13. On its keypad are five digital input buttons for numbers 1–5, a reset button R, and a display that shows the number of keys pressed since the last reset. The lock code is a sequence of four digits. At the initial state, a 0 is displayed. When a number key is pressed, the display is incremented by 1, regardless of the input. Upon the fourth key press, if the input sequence

Table 7.6 The state transition table for Problem 7.6.

Present state	Input	Next state	Output
A	0	A	0
	1	B	
B	0	E	0
	1	C	
C	0	D	1
	1	C	
D	0	A	1
	1	F	
E	0	A	0
	1	F	
F	0	E	1
	1	C	

Figure 7.13 Keypad of a digital combination lock.

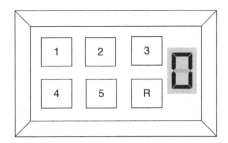

is correct, the display will change to a 0, unlocking the system. If an incorrect sequence is entered, the display will change to an E upon the fourth key press, and the user must then press the R button to reset the lock to its initial state. The user can also press the R button anytime to reset the lock. Assume that the code is set to 5152.

(1) Draw the Moore state machine of the combination lock.

(2) Draw the Mealy state machine of the combination lock.

Hint: consider the display as well as the lock status (locked, unlocked) as the output.

9 Consider a binary digital combination lock. As shown in Figure 7.14, on its keypad are two number buttons "0" and "1" and a reset button R. The length of the lock code is 6. Assume that the code is 101101. Whenever

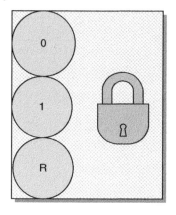

Figure 7.14 Keypad of a binary digital combination lock.

the lock detects such an input sequence, it is unlocked. For example, any one of the following binary sequences will unlock it:

101101

0101101

011010101101

1110001101101101

At any time, a press on the reset button will take the lock to its initial state. The lock enters its final state when the sequence of the code is detected. Design a state transition diagram for this lock.

10 Complete the table-based implementation of the Mealy machine for the seat belt reminder system introduced in Example 7.5.

11 Implement a C program for the state machine of the binary digital combination lock in Problem 9.
(1) Using the conditional statement approach.
(2) Using the table-based approach.

Suggestions for Reading

The concept of finite-state machine is often attributed to two physiologists Warren McCulloch and Walter Pitts for their first formal use of finite-state systems, the neural nets, in 1943, for studying the nervous activity [1]. These were later shown to be equivalent to the finite automata by Kleene [2]. Moore machine was introduced by Edward F. Moore [3]. Mealy machine was introduced by George H. Mealy [4]. The equivalence of deterministic and non-deterministic FSMs was established by Rabin and Scott [5]. A comprehensive collection of introductory materials on finite-state theories, algorithms, and the latest domain applications can be found in Ref. [6].

References

1 McCulloch, W.S. and Pitts, E. (1943) A logical calculus of the ideas immanent in nervous activity. *Bulletin of Mathematical Biology*, **5**, 115–133.

2 Kleene, S.C. (1956) *"Representation of Events in Nerve Nets and Finite Automata,"* Automata Studies, vol. **3–42**, Princeton University Press.

3 Moore, E.F. (1956) Gedanken-experiments on sequential machines, in *Automata Studies, Annals of Mathematics Studies*, vol. **34** (eds C.E. Shannon and J. McCarthy), Princeton University Press, pp. 129–153.

4 Mealy, G.H. (1955) A method for synthesizing sequential circuits. *The Bell System Technical Journal*, **34** (5), 1045–1079.

5 Rabin, M.O. and Scott, D. (1959) Finite automata and their decision problems. *IBM Journal of Research and Development*, **3**, 114–125.

6 Wang, J. (2013) *Handbook of Finite State Based Models and Applications*, CRC Press.

8

UML State Machines

The Unified Modeling Language™ (UML) is a standard graphical modeling language for the modeling, design, analysis, and implementation of software-based systems. It was first developed in the 1990s. It is the integration of notations of the Booch method, object-modeling technique (OMT), and object-oriented software engineering (OOSE). The current version of UML is UML 2.5 and was released in June 2015.

UML visualizes a software program with a collection of diagrams. These diagrams are classified into two groups: structural diagrams and behavioral diagrams. Structural diagrams emphasize the things that must be present in the system being modeled, including the following:

- *Class diagram.* A class diagram shows the structure of a system, subsystem, or component as related classes and interfaces, with their features, constraints, and relationships.
- *Package diagram.* A package diagram shows packages and relationships between them.
- *Component diagram.* A component diagram shows components and dependencies between them.
- *Composite structure diagram.* A composite structure diagram shows the internal part of a class.
- *Deployment diagram.* A deployment diagram shows the architecture of a system as deployment (distribution) of software artifacts to deployment targets.

UML behavioral diagrams describe what must occur in the system being modeled, including the following:

- *Activity diagram.* An activity diagram illustrates the dynamic nature of a system by modeling the flow of control from activity to activity.
- *Sequence diagram.* A sequence diagram describes the interactions among classes in terms of an exchange of messages over time.
- *Use-case diagram.* A use-case diagram models the functionality of a system using actors and use cases.

Real-Time Embedded Systems, First Edition. Jiacun Wang.

- *State machine diagram.* A state machine diagram describes the dynamic behavior of a system in response to external stimuli. State diagrams are especially useful in modeling reactive objects whose state changes are triggered by specific events.
- *Communication diagram.* A communication diagram describes the interactions between objects in sequence.
- *Timing diagram.* A timing diagram shows interactions when a primary purpose of the diagram is to reason about time.
- *Interaction overview diagram.* Defines interactions through a variant of activity diagrams in a way that promotes overview of the control flow.

This chapter focuses on state machine diagrams. A state machine diagram has four types of elements, namely states, transitions, events, and actions. States represent the possible operational modes of an object. Transitions represent legal changes from one state to another. Events are the labels of a transition, defining under what conditions the transition occurs. Actions are the activities that take place within a state. This chapter introduces the fundamental concepts of UML state machines, such as hierarchy, concurrency, and modularity of graphic representation, as well as various graphic elements. At the end of the chapter, the antilock braking system (ABS) is used as example to show the application of UML state machines in real-world embedded system behavior modeling.

8.1 States

A state in a UML state machine is modeled using a circular rectangle with the state's name being represented either inside the rectangle or outside the rectangle using a name tab. Figure 8.1 shows three different ways of representing a state called *On*. In the representation shown in Figure 8.1a, the state name is inside the rectangle. This type of representation is usually used to model simple states. In Figure 8.1b, the state name is inside a name tab. This type of state representation is useful when modeling orthogonal composite states. In Figure 8.1c, the rectangle is divided into two different compartments. The state name is given inside the first compartment. This type of representation is usually useful when modeling simple dynamic states or hierarchical composite states.

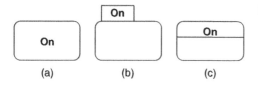

Figure 8.1 Representations of a state.

(a)   (b)   (c)

Figure 8.2 Initial state marker and end state marker.

Start   End

In a state diagram, there is an *initial state marker*, represented by a big solid dot, which represents the starting point of a machine. It has neither real existence nor incoming transition. There may also be *end state markers*, represented by a big dot with a circle around it, which indicate the end of processing. Both initial state markers and end state markers are *pseudostates* in UML state machines, which will be discussed in detail later. Figure 8.2 shows the start marker and end marker. Not all systems have an end marker.

Three types of states are defined in UML state machines: *simple states*, *composite states*, and *submachine states*. A simple state represents a basic situation, and it does not have substates – it contains neither *regions* nor submachine states. A composite state has substates and contains one or more regions. A submachine state specifies the insertion of the specification of a submachine state machine.

Example 8.1 *UML State Machine for a Landline Telephone*
Consider as an example of the behavioral representation of a landline telephone, which is shown in Figure 8.3. It is a high-level model and has only two states: *Idle* and *Active*, where the *Idle* state indicates that the phone is not in use while the *Active* state indicates that the phone is in use. Events that cause state changes are *pick up* and *hang up*.

States may be dynamic, that is, they can perform some type of actions while they are active. At the time of execution of an action, the state will not accept any event until the action is completed. UML state machines support three types of behaviors within a state:

- *entry*: the specified behavior is executed at the moment the state becomes active.
- *exit*: the specified behavior is executed at the moment the state becomes inactive.
- *do*: the specified behavior is executed continuously as long as the state is active.

These behaviors can be specified in a compartment within a state.

Figure 8.3 UML state machine for a landline telephone.

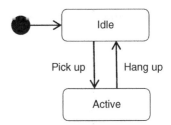

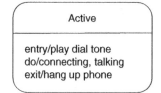

Figure 8.4 A simple state with entry, do, and exit behaviors specified.

Example 8.2 *State with Compartments*

Figure 8.4 shows the telephone *Active* state with two compartments. The upper compartment has the state name, while the lower compartment specifies the state behaviors.

8.2 Transitions

States represent the modes of operation that we are interested in while modeling the behavior of an object. Transitions, on the other hand, are used to depict the changes of modes. A transition is graphically represented as a directed arc from a source state to a target state. Figure 8.5 shows the general representation of a transition. A transition can be optionally labeled by the following elements:

- *Event-list*: an event or a list of events that cause the transition to be triggered or fired. In case there is a list of events, the occurrence of one event from this list is enough to trigger the transition.
- *Guard*: a conditional statement that is evaluated to determine whether the transition should be triggered or not. A guard can be a variable range specification, such as [speed > 60], or some other constraint must be met.
- *Action-list*: an operation or set of operations that will execute when the transition is triggered. Actions normally take a short period of time to perform. They may occur when a transition is taken, when a state is entered, or when a state is exited.

During the execution of an object or a system, a particular state can be active or inactive. The occurrence of a transition indicates the following:

- The source state was the active state before the triggering of the transition.
- The source state exits because of the transition occurrence and becomes inactive.
- The target state becomes the active state.

Figure 8.5 A transition.

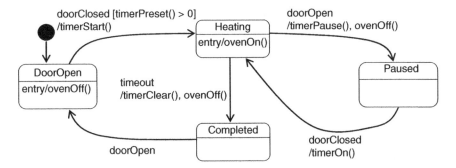

Figure 8.6 A simplified UML state machine for an oven.

Example 8.3 *A Simplified State Machine for an Oven*
Figure 8.6 depicts a simplified UML state machine for an oven. There are four states in this model: *DoorOpen, Heating, Paused,* and *Completed.* Consider the transition from *DoorOpen* to *Heating.* The event that causes the transition is *doorClosed.* The guard is *timePreset() > 0,* meaning that you have to set the timer to a value that is greater than 0 in order for the oven to start heating. When this transition is triggered, the timer is armed, which is represented by the action *timerStart().*

8.3 Events

Events are internal or external interactions that cause transitions in the behavior of an object or a system. For example, the state machine shown in Figure 8.6 has two events: *doorOpen* and *doorClosed.* The response of the system to an event depends on its current state and the type of events this state accepts. For example, the state machine shown in Figure 8.6 shows that, while the oven is in the *Heating* state, the event of *doorOpen* changes the oven state to *Paused,* indicating that the heating is paused. If an event occurs while the system is currently in a state that cannot handle this event, then the system behavior will not change and the event will be ignored. The two states *DoorOpen* and *Paused* can only accept the event *doorClosed,* while the states *Heating* and *Completed* only respond to the event *doorOpen.*

UML defines four kinds of events:

Signal events. A signal event is a named object that is dispatched asynchronously by one object and received by another.

Call events. A call event represents the dispatch of an operation. It is a synchronous event. The general format of call events is

```
event_name(parameters).
```

Parameters are optional. The call events shown in Figure 8.6 do not have parameters.

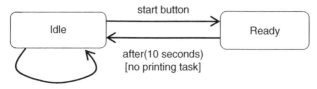

when (11:59 PM)/self test

Figure 8.7 Change event and time event with a printer.

Change events. A change event represents a change in the state of the satisfaction of some condition. The general format of this type of events is

when(condition).

Time events. A time event represents the passage of time. This type of event keeps track of the time that a state is active and compares it to a boundary value. An event occurs when the boundary value is exceeded. The general format of this type of events is

after(time),

where time can be either absolute or relative. When an object enters a state, any time-out function from that state is started. When a time-out expires, the state machine receives the expiration as an event. When an object leaves a state, any time-out that was started upon entry to that state is cancelled. Only one time-out can be used per state.

Figure 8.7 illustrates a change event and a time event. It is a partial representation of the behavior of a printer. The printer is in the *Idle* state when it is powered on but not printing. The printer does a self-test every night at 11:59 PM if it is idle. When the start button is pressed, the printer enters the *Ready* state, in which it is ready to print. However, if it does not receive any printing task from clients within 10 seconds, the printer returns to the *Idle* state.

8.4 Composite States

The most distinguished extension of UML state machines to the traditional FSM formalism is the introduction of composite states. A composite state has substates and contains one or more regions. Composite states can be specified in hierarchy or orthogonality. In case of hierarchy, the composite state is composed of one region. As for orthogonality, a composite state encloses two or more regions, where each region executes concurrently. A region in a hierarchical or orthogonal state contains states and transitions. The states inside a region in turn can also be simple states, composite states, or submachine states.

8.4.1 Hierarchy

Hierarchy is introduced by allowing a state to encompass other states. Hierarchical states represent abstraction of the software behavior, where lower level abstractions are depicted as states that are enclosed within other states. This is a traditional way of handling complexity. Hierarchical states are an ideal mechanism for hiding internal details because the designer can easily zoom out or zoom in to hide or show nested states, respectively.

Example 8.4 *Decomposition of the* **Active** *State Shown in Figure 8.3*
Recall that Figure 8.3 depicts the phone behavior at a high level of abstraction. To analyze and understand the behavior of the phone while it is being used, it is necessary to develop a more detailed representation, or a lower level abstraction, of what is happening while the phone is on. This in-depth representation of the operational phone behavior should include multiple modes of interest, such as dialing, connecting, and talking. Therefore, there is a need to refine the *Active* state.

Figure 8.8 shows the representation of the phone behavior while it is active as a composite state that supports hierarchy. This composite state consists of 6 simple states and 10 transitions. The *Ready* state is the default state to enter when the composite state is activated, and it indicates that the phone is ready and can be used to make a call. The action of playing a dial tone is performed immediately upon entry to the state. The moment this state exits, due to a *dialDigit* event, the dial tone should be stopped. If the user does not dial a digit in 10 seconds, the phone enters the *Not Ready* state, and then, the user can press

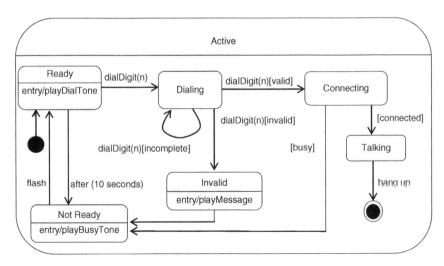

Figure 8.8 Partial representation of the *Active* composite state.

the flash button to return the phone to the *Ready* state. When the user starts to dial, the phone enters the *Dialing* state, and the user has to dial all the required number of digits to complete the dialing. If the dialed number is a valid phone number, the phone enters the *Connecting* state; otherwise, it enters the *Invalid* state and plays an error message. After the error message is played, it changes to the *Not Ready* state. If the called party's line is in use, the phone changes from the *Connecting* state to the *Not Ready* state; otherwise, it moves the *Talking* state. When the caller hangs up, the *Active* state becomes inactive.

We can further decompose a substate's behavior into a lower level abstraction and represent it as a composite state. For example, we can replace the *Connecting* state shown in Figure 8.8 with a composite state that describes the connecting behavior in a finer granularity as shown in Figure 8.9.

We introduced a new symbol in Figure 8.9, which is the diamond representing the conditional branch, called the *choice pseudostate*. It evaluates the guards of the triggers of its outgoing transitions to select only one outgoing transition. In the state machine shown in Figure 8.8, we can also use a choice pseudostate to branch out from the state *Dialing* to the states *Dialing*, *Connecting*, and *Invalid*, to make the condition check more intuitive.

Drawing the transitions that end on the board of a composite state looks less nice. Later on, the entry/exit pseudostates will be introduced to make the graphical representation more structured.

When there are a large number of states in a state machine, it is desirable to hide the decomposition of a composite state and represent it with a simple state graphic, so that all states can fit into the graphical space available for the diagram. To differentiate the composite state from a simple state in graphic representation, we use a special "composite" icon, usually in the lower right-hand corner. This icon, consisting of two horizontally placed and connected states,

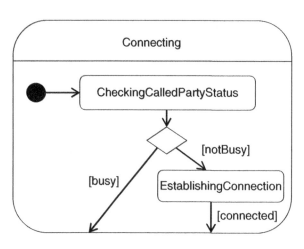

Figure 8.9 Decomposition of the *Connecting* state shown in Figure 8.8.

Figure 8.10 Composite state *Connecting* with decomposition hidden.

is an optional visual cue that the state has a decomposition that is not shown in this particular diagram. Instead, the contents of the composite state are described in a separate diagram. For example, the composite state *Connecting* state in the state machine shown in Figure 8.8 can be replaced with the one shown in Figure 8.10.

8.4.2 Orthogonality

Orthogonal states represent concurrent behavior. An orthogonal state has more than one region. Regions of parallel states are separated by dashed lines. Each dashed line divides the composite state into two separate regions. These orthogonal regions run concurrently. In other words, the software system or subsystem must be in *all* its regions the moment the composite state becomes active. Communication between orthogonal parts can be achieved through signal events and/or call events. This notion of orthogonality is very helpful in representing subcomponents of a system and pertaining modularity in the behavioral representation.

Example 8.5 *UML State Machine Diagram of a Soda Machine*
Consider a soda vending machine. We are interested in its behavior when a customer makes a purchase. Its high-level model is depicted in Figure 8.11, in which the state *Dispensing* is a composite state, but is represented as a simple state with a composite state icon.

Internally, the state *Dispensing* has two paths of operations that run in parallel. One path handles the soda delivery, and the other path does the

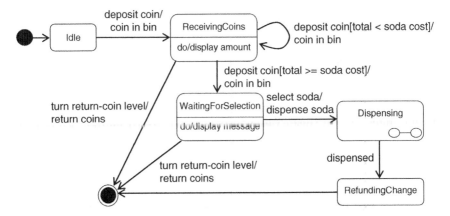

Figure 8.11 UML state machine of a soda machine.

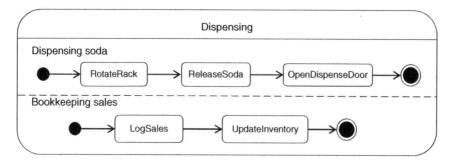

Figure 8.12 Orthogonal composite state *Dispensing*.

bookkeeping of sales. Figure 8.12 shows the behavior of the *Dispensing* state. It has two orthogonal regions.

8.4.3 Submachine States

A submachine is a state machine inserted as a state in another state machine. The same submachine can be inserted more than once. Submachine states are semantically equivalent to composite states in that they are made up of internal states and transitions. However, submachine states provide a means to encapsulate states and transitions so that they can be reused. The name compartment of a submachine state holds the (optional) name of the state, as a string. The name of the referenced state machine is shown as a string following ":" after the name of the state. For example, to follow the submachine state name convention, we can rename the composite state *Dispensing* in Figure 8.11 to *Disp:Dispensing*, where *Disp* is the new name of the state, and the whole name indicates that a submachine state machine named *Dispensing* is to be inserted here.

8.5 Pseudostates

In addition to simple states and composite states, UML state machines also defined a set of pseudostates that are used to precisely specify the dynamic behavior of a system. We have introduced the initial pseudostates and choice pseudostates before. In this section, we introduce a few other important pseudostates. They are history pseudostates, entry/exit pseudostates, fork/join pseudostates, and terminate pseudostates.

8.5.1 History Pseudostates

In some systems, it is relevant to remember the last active internal state of a composite state or a submachine state when this composite state turns inactive. For example, when we open the door of an oven while it is heating, the state *Heating* is remembered by the system. Later on, when the door is closed, the oven resumes to heat. UML state machines support this type of behavior through the use of *history pseudostates*. UML defined two kinds of history, *shallow history* and *deep history*. A shallow history pseudostate, shown as a circled letter H, is used to remember the last active internal state of a composite state or a submachine state but not the substates of that last active internal state. A transition coming into the shallow history vertex is equivalent to a transition coming into the most recent active substate of a state. At most, one transition may originate from the history connector to the default shallow history state. This transition is taken in case the composite state had never been active before, because there is simply no history. A deep history is a shallow history recursively reactivating the substates of the most recent active substate. It is represented as the shallow history with a star (H* inside a circle).

Figure 8.13 shows two identical state machines. The only difference between them is that one uses a shallow history and the other uses a deep history in the composite state *State*1. At the highest level, each machine has two states: *State*1 and *State*2. Within *State*1, there are two substates: *State*11 and *State*12, where *State*12 is also a composite state, which is made up of two substates: *State*121 and *State*122. Suppose that each machine transitions out of *State*1 while it is at *State*122 to *State*2 and then transitions back to *State*1. With the shallow history pseudostate, the first machine will return to *State*12 and start with *State*121. On the other hand, the second machine will directly return to *State*122, because the deep history pseudostate remembers the leaf substate within *State*1.

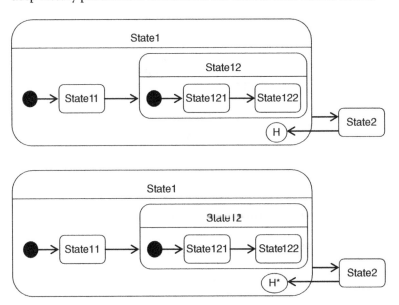

Figure 8.13 Shallow history pseudostate and deep history pseudostate.

Example 8.6   *State Machine of a CD Player*

Consider a CD player with high-level behavior depicted by the state machine shown in Figure 8.14a. The *Playing* state is a composite state, and its internal behavior is specified in Figure 8.14b, which shows how the CD player plays back the three songs stored in a CD according to the manual selection by a user. Of course, this is a simplified model. The intention here is to show the use of shallow history pseudostates. Notice the shallow history pseudostate icon in the composite state representation in Figure 8.14b.

The state machine starts with the *Stopped* state. When the *press play* event occurs for the first time, the state machine changes to the *Playing* state. Because this is the first time that the *Playing* state becomes active, this composite state will start with the substate *Song*1, and the playing behavior changes among the *Song*1, *Song*2, and *Song*3 states, depending on the events *press next* and *press prev*. If the *press stop* or *press pause* event occurs, the machine leaves the *Playing* state, and meanwhile, the last active state is remembered by the shallow history pseudostate. When the *press play* event occurs again, the *Playing* state is reactivated, and internally, the machine restores the substate represented by the shallow history pseudostate; thus, the interrupted song is played.

Because no single internal state of the state *Playing* has substates, there is no difference between using a shallow history pseudostate and using a deep history pseudostate in this example.

8.5.2   Entry and Exit Points

When a composite state or submachine state becomes active, the internal substate to enter is the initial state by default for each region, or the last active state

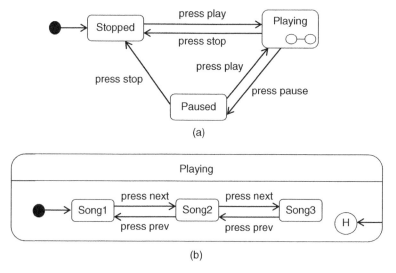

Figure 8.14 UML state machine of a CD player. (a) High-level state machine. (b) Composite state *Playing*.

in case there is a history pseudostate and the composite state or submachine state was active before. Sometimes, we may not want to enter a submachine at the default state. Instead, we want to enter a particular internal state. In that case, an entry point pseudostate can be used. An *entry point pseudostate* is an entry point of a state machine or composite state. In each region of the state machine or composite state, it can have at most one single transition to a vertex within the same region. An entry point is shown as a small circle on the border of the state machine diagram or composite state, with the name associated with it.

By default, a state machine exits a composite state when all internal substates of the composite state become inactive. In a similar manner to entry points, it is possible to have named alternative exit points. An *exit point pseudostate* is an exit point of a state machine or composite state. Entering an exit point within any region of the composite state implies the exit of this composite state or submachine state. It also implies the triggering of the transition that has this exit point as source in the state machine enclosing the composite state. An exit point is shown as a small circle with a cross on the border of the state machine diagram or composite state, with the name associated with it.

Example 8.7 *A Data Processing State Machine*
Figure 8.15 shows a simple state machine for data processing. It is composed of four high-level states, namely *Reading Data, Processing Data, Displaying Results,* and *Reporting Error. Processing Data* is a composite state and has two substates: *Sorting* and *Processing*. By default, data are sorted first and then get processed. If the data are already sorted, then the sorting phase is skipped. In the machine, the unsorted transition goes to the default initial state, *Sorting,* while the sorted transition connects to the *Processing* state via the entry point

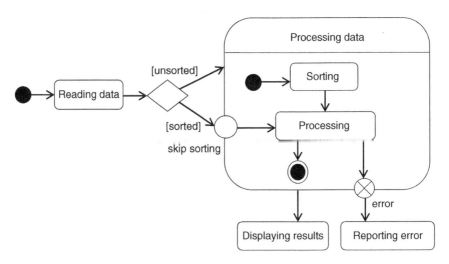

Figure 8.15 A simple state machine for data processing.

skip sorting. If the *Processing* state is finished properly, the machine transitions to the *Displaying Results* state by default. If the state terminates with an error, the machine moves to the *Reporting Error* state via the exit point *error*.

8.5.3 Fork and Join Pseudostates

A *fork pseudostate* splits an incoming transition into two or more transitions terminating on target states that are in different orthogonal regions of a composite state. The transition outgoing from a fork pseudostate must not have guards or triggers because it is unconditional by definition. On the contrary, *a join pseudostate* merges several transitions originating from source states in different orthogonal regions of a composite state. The transition entering a join vertex cannot have guards or triggers. The notation for a fork and join pseudostate is a short heavy bar and is illustrated in Figure 8.16.

8.5.4 Terminate Pseudostates

A *terminate pseudostate* is used to represent a complete stoppage of the behavior of a state machine. This implies that the execution of the system has been terminated. At the moment of termination, the state machine cannot respond to events anymore and, as a result, cannot change its behavior. An example of a complete stoppage to the CD player behavior is that it runs out of battery while it is playing. Figure 8.17 shows a modified behavior to the CD player example. The terminate pseudostate is shown in the figure as a cross symbol.

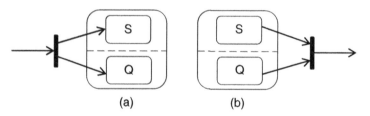

(a) (b)

Figure 8.16 Fork and join pseudostates. (a) Fork splits a transition. (b) Join merges transitions.

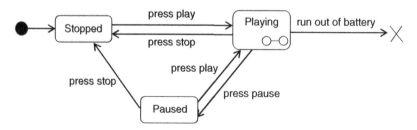

Figure 8.17 CD player state machine with a terminate pseudostate.

8.6   UML State Machine of Antilock Braking System

The notion of hierarchy and orthogonality in composite states and submachine states gives UML state machine diagrams the power to support hierarchy, concurrency, and modularity in system behavior modeling. It allows software practitioners to model a system at multiple abstraction levels and design each component independently. At the highest level of abstraction, we consider components (or subsystems) of a system and how these components interact with each other. Then, we model each component. A component can be further decomposed into a set of subcomponents. This kind of refinement process can go on and on. When a component reaches a desired level of abstraction, we identify all states the component can go in and then consider state transitions, events that trigger these transitions, and the system actions after the transitions. This section discusses the hierarchical modeling idea with the ABS that is introduced in Chapter 1.

Recall that sensors, valves, pumps, and an electrical control unit (ECU) are the major functional components in an ABS. Critical sensors that detect a wheel lockup are wheel speed sensors and deceleration sensors. Valves include isolation valves and dump valves. These components operate concurrently. Thus, the state machine of the ABS should have six components, represented by six regions in Figure 8.18. The ECU is the brain of the ABS. All other components act upon commands sent by the ECU. The communication among components should be reflected in the state machine model of each component. Next, we create the state machine diagram for each component and discuss how these state machines interact with each other.

Figure 8.19 shows the wheel speed sensor state machine. It has only one state named *Idle*. A *get_wheel_speed* event, which is triggered by a signal

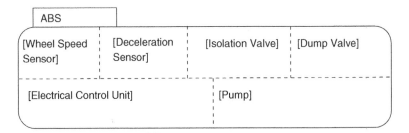

Figure 8.18 Components of ABS.

Figure 8.19 Wheel speed sensor state machine.

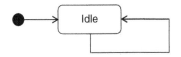

get_wheel_speed()/return_wheel_speed()

sent from the ECU, results in a transition from the *Idle* state to itself, and a *return_wheel_speed*() message is sent back to the ECU. The state machine of the deceleration sensor is identical to the wheel speed sensor model, with the only difference being the label of the self-transition, which is *get_deceleration*()/*return_deceleration*().

Both the isolation valve and pump valve have two states: *Closed* and *Open*. Their state machine diagrams are exactly the same, as shown in Figure 8.20. They are initially at the *Closed* state. When a valve receives an *open* signal from the ECU, it switches to the *Open* state. Then, a *close* signal from the ECU triggers the valve to change from the *Closed* state to the *Open* state.

The pump has two states: *Idle* and *Pumping*, with *Idle* being its entry state. A *pumping* signal from the ECU will trigger the pump to move to the *Pumping* state. Then, a *stop* signal from the ECU will change the pump from the *Pumping* state to the *Idle* state. The state machine is shown in Figure 8.21.

At the highest level, the ECU state machine has two states: *On* and *Off*, where *Off* is the entry state and *On* is a composite state, as shown in Figure 8.22. When the ECU is powered on, it first enters the *Initializing* state. An important

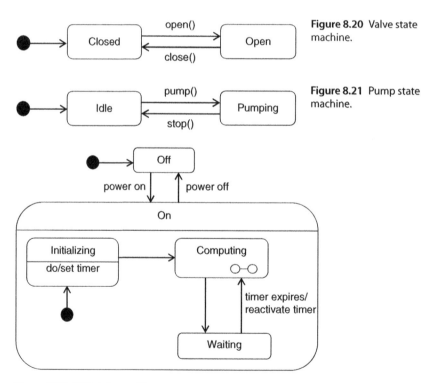

Figure 8.20 Valve state machine.

Figure 8.21 Pump state machine.

Figure 8.22 ECU state machine.

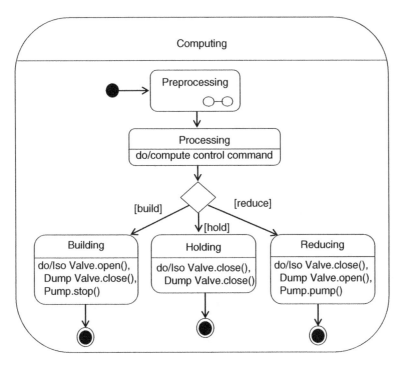

Figure 8.23 Composite state *Computing*.

internal event at this state is setting and activating the timer. The ECU computes the control command periodically. When the control unit is initialized, it moves to the *Computing* state. This is a composite state, and its internal behavior is described in Figure 8.23. When the ECU exits the *Computing* state, it enters the *Waiting* state, waiting for the beginning of the next control cycle. When the timer expires, it triggers the transition to the *Computing* state, and meanwhile, the timer is reactivated to clock the next control cycle.

The entry state of the *Computing* state is the *Proprocessing* state, within which sensor data are read and converted to digital words. The *Preprocessing* state shown in Figure 8.23 is a composite state, and its internal behavior is presented in Figure 8.24. After sensor data are preprocessed, the ECU enters the *Process-ing* state and starts to compute the control command. The result could be build the braking pressure, hold the braking pressure, or reduce the braking pressure. So, the *Processing* state transitions to one of the three states: *Building*, *Holding*, and *Reducing*, depending on the computation result. At the *Building* state, the ECU sends an *open* signal to the isolation valve and a *close* signal to the dump valve. At the *Holding* state, the ECU sends an *close* signal to the isolation valve and a *close* signal to the dump valve. At the *Reducing* state, the ECU sends a *close* signal to the isolation valve and an *open* signal to the dump valves. In this diagram, the convention

```
target_component_name.signal_name(parameter)
```

is used for intercomponent communication. For example, the event *Iso Valve.open*() means that a signal *open* is sent to the isolation valve component. A component is a region of a state machine.

Figure 8.24 describes the internal behavior of the *Preprocessing* composite state. The state has two regions, one for the wheel speed sensor data preprocessing and the other for the deceleration sensor data preprocessing. Each substate machine has two states: *Reading* and *A/D Converting*. In the *Reading* state, the ECU sends a reading signal to the corresponding sensor to obtain the measurement. Readings are converted to digital words in the *A/D Converting* state.

In summary, the ABS UML state machine has six regions, each region for a major component in the system. The substate machines of all components except the ECU are rather simple. The state machine of each sensor has only one state, while the state machine of each valve has two states. The state machine of the pump also has only two states. These components act upon signals of command from the ECU. The dynamic behavior of the ECU is specified at three levels of abstraction. The events that occur upon the ECU sending commands to all other components are reflected in the transition labels.

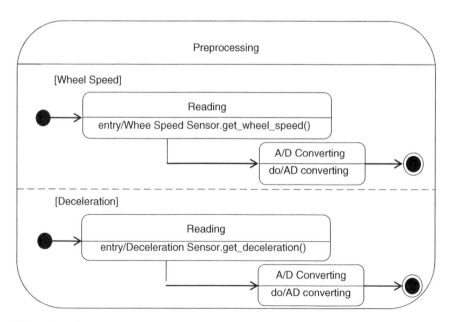

Figure 8.24 Composite state *Proprocessing*.

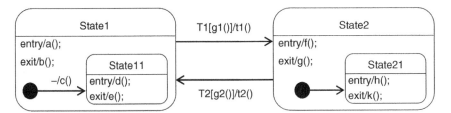

Figure 8.25 A state machine for Problem 1.

Exercises

1 Consider the state machine shown in Figure 8.25.
   (1) When transition T1 occurs, what actions will be followed in order?
   (2) When transition T2 occurs, what actions will be followed in order?

2 Develop a UML state machine to specify the dynamic behavior of a door that can be opened or closed. When it is closed, it can be locked and unlocked. Notice that you can open or close a door only if the doorway is cleared.
   (1) Use only simple states.
   (2) Use a composite state *Closed* to model the behavior of the door when it is closed. The composite state has two internal states: *Unlocked* and *Locked*.

3 Draw the state machine for a luggage belt system. The belt is started when the start button is pressed and runs either until the stop button is pressed or until there is no luggage on the belt. This no-luggage condition prevails when no luggage has been detected in the previous 60 seconds.

4 Draw the state machine for a simple battery charger that can charge two batteries in parallel. The charger has three modes: idle, discharging, and charging. The charger starts charging the batteries when a button is pushed. However, before each battery can be charged, it must be discharged. When each battery is discharged, it sends a signal to the charger to change the mode to charging. When a battery is charged, it sends notification to the charger. When both batteries are charged, the charger returns to the idle mode.

5 Table 8.1 lists the power modes of an LCD projector, in which its current power mode is indicated by the status of an indicator in the projector. Power status can be changed by pressing the power button in a remote control. For example, when the power is off, a press on the button will turn it on. When the power is on, a press on the button will switch the power off.

Table 8.1 Power modes of a projector.

Indicator status	Explanation
Red, lit	Power is off (standby mode). Press power button to start projection
Red, flashing	Power is off (standby mode), and the Power On Blink feature is set to yes
Green, lit	It is projecting
Green, flashing	It is preparing to project. Flashing lasts for 5 seconds
Orange, lit	It is preparing to switch power off. Flashing lasts for 5 seconds
Orange, flashing	Power is pressed again in the power-off preparation mode
No illumination	Main power is switched off

In addition, when the power is on, if the projector receives no signal from the remote control upon a press for 5 minutes, it will behave as if having received a power button signal. Develop a UML state machine to describe the dynamic behavior of the power status of the LCD projector.

6 To receive a driver license, an adult must take and pass a written permit test and a road test. If one fails the written permit test, he has to wait for at least 1 week to take it again. When he passes the written permit test, he can take the road test 3 months later. If he passes the test, he will receive his driver license. If he fails, he has to wait for a minimum of 2 weeks to take the test again. If he fails the road test three or more times, he has to wait for at least 6 months before he can take the test again. Draw a state diagram to specify the testing process. Consider only four states: *Permit Testing*, *Waiting for Permit Testing*, *Road Testing*, and *Waiting for Road Testing*. The guard of "wait for at least one week" can be specified by a change event as follows:

```
when(waiting_time ≥ 1 week)
```

7 Consider the wall-mounted control unit of a garage door opener. The control unit has two buttons: a door button that opens or closes the garage door, depending on the door's current state, and a light button that switches the light on the motor unit on or off. Whenever the door button is pressed, it will turn the light on as well as open or close the door. When the light is left on for 60 seconds, it will turn off automatically. Create a UML state machine for the behavior of the control unit, door, and light.

8 Create a UML state machine that models the operation of a simple cell phone according to the following specification:

- The cell phone has an On/Off switch.
- It has a numeric keypad that produces a keypad press event with a digit as its argument.
- The phone has a three-way switch that is set to ringing, vibrating, or both; it determines the action of the phone when a call comes in.
- The phone has an action button that
  - initiates a call when seven digits have been entered,
  - answers a call when the phone is ringing or vibrating, and
  - terminates a call (hangs up) if a call is in progress.
- If the action button is pressed when fewer than seven digits have been entered, the digits are erased.
- When dialing, the interval between any two consecutive digits pressed cannot exceed 10 seconds; otherwise, a time-out event will terminate the call.
- Finally, the phone has a display that shows the digits that have been pressed so far, if any.

Use composite states as appropriate in your model to increase its readability.

Suggestions for Reading

UML state machine diagrams originated from Harel Statecharts [1]. David Harel described how the language of Statecharts came into being in [2]. UML state machine diagrams are part of the OMG UML. The specification of the latest version of UML 2.5 is downloadable from the OMG official website [3, 4].

It is possible to directly generate source code from state machines to automate the design process. For example, IBM Rational allows C, C++, Java, or Ada code generation from UML state machines [5]. Samek [6] gives a detailed description of how to generate C/C++ code from state machine through numerous examples of embedded systems.

References

1 Harel, D. (1987) Statecharts: a visual formalism for complex systems. *Science of Computer Programming*, **8** (3), 231–274.

2 Harel, D. (2009) Statecharts in the making: a personal account. *Communications of the ACM*, **52** (3), 67–75.

3 Dennis, A., Wixom, B.H., and Tegarden, D. (2015) *Systems Analysis and Design: An Object-Oriented Approach with UML*, 5th edn, Wiley.

4 OMG OMG Unified Modeling Language™ (OMG UML), http://www.omg.org/spec/UML/2.5, Version 2.5, 2015 (accessed 21 March, 2017).

5 IBM IBM Rational Software, http://www-01.ibm.com/software/rational/ (accessed 21 March, 2017).

6 Samek, M. (2008) *Practical UML Statecharts in C/C++: Event-Driven Programming for Embedded Systems*, 2nd edn, Newnes, Newton.

9

Timed Petri Nets

Petri nets were introduced in 1962 by Dr Carl Adam Petri. Petri nets are a powerful modeling formalism in computer science, system engineering, and many other disciplines. Petri nets combine a well-defined mathematical model with a graphical representation of the dynamic behavior of discrete event-driven systems. The theoretical aspect of Petri nets allows precise modeling and analysis of system behavior, while the graphical aspect enables visualization of the state changes of the modeled system. This combination is the main reason for the great success of Petri nets. Consequently, Petri nets have been used to model various kinds of event-driven systems such as embedded systems, communication systems, manufacturing plants, command and control systems, real-time computing systems, logistic networks, and workflows, to mention only a few important examples. Timed Petri nets, in which job execution times or event durations are specified, are able to catch the time-related performance or real-time properties of a system.

9.1 Petri Net Definition

A Petri net is a particular kind of bipartite directed graph, populated by four types of objects: *places, transitions, directed arcs,* and *tokens.* Directed arcs connect places to transitions or transitions to places. In its simplest form, a Petri net can be represented by a transition together with an input place and an output place. This elementary net may be used to represent various aspects of the modeled systems. For example, a transition and its input place and output place can be used to represent a data processing event, its input data and output data, respectively, in a data processing system. In order to study a system's dynamic behavior, in terms of states and state changes, using Petri nets, each place may potentially hold either none or a positive number of tokens. Tokens are a primitive concept for Petri nets in addition to places and transitions. The presence or absence of a token in a place can indicate whether a condition associated with this place is true or false, for instance.

Real-Time Embedded Systems, First Edition. Jiacun Wang.
© 2017 John Wiley & Sons, Inc. Published 2017 by John Wiley & Sons, Inc.

A Petri net is formally defined as a five-tuple $N = (P, T, I, O, M_0)$, where

- $P = \{p_1, p_2, \ldots, p_m\}$ is a finite set of places;
- $T = \{t_1, t_2, \ldots, t_n\}$ is a finite set of transitions, $P \cup T \neq \emptyset$, and $P \cap T = \emptyset$;
- $I: T \times P \to N$ is an *input function* that defines directed arcs from places to transitions, where N is a set of all nonnegative integers;
- $O: T \times P \to N$ is an *output function* that defines directed arcs from transitions to places; and
- $M_0: P \to N$ is the *initial marking*.

A *marking* in a Petri net is an assignment of tokens to the places of a Petri net. Tokens reside in the places of a Petri net. The number and location of tokens may change during the execution of a Petri net. The tokens are used to define the execution of a Petri net. A place containing one or more tokens is said to be *marked*.

Most theoretical work on Petri nets is based on the formal definition of Petri nets. However, a graphical representation of a Petri net is much more useful for illustrating the structure and dynamics of the modeled system. A Petri net graph is a Petri net depicted as a bipartite directed multigraph. Corresponding to the definition of Petri nets, a Petri net graph has two types of nodes: a circle represents a place, and a bar or box represents a transition. Directed arcs (arrows) connect places and transitions, with some arcs directed from places to transitions and other arcs directed from transitions to places. An arc directed from a place p_j to a transition t_i indicates that p_j is an *input place* of t_i, denoted by $I(t_i, p_j) = 1$. An arc directed from a transition t_i to a place p_j indicates that p_j is an *output place* of t_i, denoted by $O(t_i, p_j) = 1$. If $I(t_i, p_j) = k$ (or $O(t_i, p_j) = k$) for some $k > 1$, then there exist k directed (parallel) arcs connecting place p_j to transition t_i (or connecting transition t_i to place p_j). Usually, in the graphical representation, parallel arcs connecting a place (transition) to a transition (place) are represented by a single directed arc labeled with its multiplicity or weight k. A circle containing a dot represents a place containing a token.

Example 9.1 *A Simple Petri Net*
Figure 9.1 shows a simple Petri net. It has four places and three transitions. At the initial marking, the place p_1 has two tokens, while all other places have no

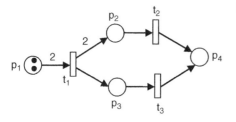

Figure 9.1 A simple Petri net.

tokens. Its five tuples are as follows:

$$P = \{p_1, p_2, p_3, p_4\};$$

$$T = \{t_1, t_2, t_3\};$$

$$I(t_1, p_1) = 2, \quad I(t_1, p_i) = 0, \quad i = 2, \ 3,4;$$
$$I(t_2, p_2) = 1, \quad I(t_2, p_i) = 0, \quad i = 1, \ 3,4;$$
$$I(t_3, p_3) = 1, \quad I(t_3, p_i) = 0, \quad i = 1, \ 2,4;$$

$$O(t_1, p_2) = 2, \quad O(t_1, p_3) = 1, \quad O(t_1, p_i) = 0, \quad i = 1,4;$$
$$O(t_2, p_4) = 1, \quad O(t_2, p_i) = 0, \quad i = 1, 2,3;$$
$$O(t_3, p_4) = 1, \quad O(t_3, p_i) = 0, \quad i = 1, 2,3;$$

$$M_0 = (2, \ 0, \ 0, \ 0).$$

In the initial marking, p_1 is the only marked place. Notice that there is a label of 2 on the arc from p_1 to t_1 and from t_1 to p_2. It means that the weights of these two arcs are 2: when t_1 fires, two tokens in p_1 will be taken away and two tokens will be placed in p_2.

9.1.1 Transition Firing

The execution of a Petri net is controlled by the number and distribution of tokens in the Petri net. By changing the distribution of tokens in places, which may reflect the occurrence of events or execution of operations, for instance, one can study the dynamic behavior of the modeled system. A Petri net is executed by *firing* transitions. Denote the number of tokens in a place p in the marking M by $M(p)$. We now introduce the enabling rule and firing rule of a transition, which govern the flows of tokens:

- *Enabling Rule*: A transition t is said to be *enabled* if each input place p of t contains at least the number of tokens equal to the weight of the directed arc connecting p to t, that is, $M(p) \geq I(t, p)$ for all p in P. If $I(t, p) = 0$, then t and p are not connected, so we don't care about the marking of p when considering the firing of t
- *Firing Rule*: Only enabled transitions can fire. The firing of an enabled transition t removes the number of tokens equal to $I(t, p)$ from each input place p and deposits the number of tokens equal to $O(t, p)$ in each output place p.

Mathematically, firing t in M yields a new marking

$$M'(p) = M(p) - I(t,p) + O(t,p) \quad \forall p \in P.$$

Notice that since only enabled transitions can fire, the number of tokens in each place always remains nonnegative when a transition is fired. Firing a transition can never try to remove a token that is not there.

A transition without any input place is called a *source transition,* and one without any output place is called a *sink transition.* Note that a source transition is unconditionally enabled and that the firing of a sink transition consumes tokens, but doesn't produce tokens.

A pair of place p and transition t is called a *self-loop,* if p is both an input place and an output place of t. A Petri net is said to be *pure* if it contains no self-loops.

Example 9.2 *Transition Firing*

Consider the simple Petri net shown in Figure 9.1. In the initial marking M_0, only t_1 is enabled. Firing of t_1 results in a new marking, say M_1. It follows from the firing rule that

$$M_1 = (0,\ 2,\ 1,\ 0).$$

The new token distribution of this Petri net is shown in Figure 9.2. Then in the marking M_1, both transitions t_2 and t_3 are enabled. If t_2 fires, the new marking, say M_2, is

$$M_2 = (0,\ 1,\ 1,\ 1).$$

If t_3 fires, the new marking, say M_3, is

$$M_3 = (0,\ 2,\ 0,\ 1).$$

9.1.2 Modeling Power

The typical characteristics exhibited by the activities in an embedded system, such as concurrency, decision-making, synchronization, and priorities, can be modeled effectively by Petri nets.

1. *Sequential execution.* In Figure 9.3, the transition t_2 can fire only after the firing of t_1. This imposes the precedence constraint "t_2 after t_1." Such precedence constraints are typical among tasks in an embedded system. Also, this Petri net construct models the causal relationship among activities.

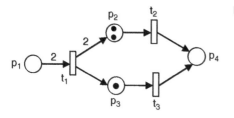

Figure 9.2 Firing of transition t_1.

Figure 9.3 Two sequential transitions.

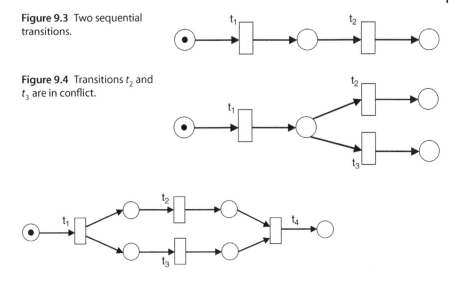

Figure 9.4 Transitions t_2 and t_3 are in conflict.

Figure 9.5 Transitions t_2 and t_3 are concurrent. Transition t_4 synchronizes two sequences.

2. *Conflict.* Transitions t_2 and t_3 are in conflict in Figure 9.4. Both are enabled; however, the firing of any transition leads to the disabling of the other transition. Such a situation will arise, for example, when two tasks compete for the CPU or any other shared resource. The resulting conflict may be resolved in a purely nondeterministic way or in a probabilistic way, by assigning appropriate probabilities to the conflicting transitions.
3. *Concurrency.* In Figure 9.5, t_2 and t_3 are concurrent. Concurrency is an important attribute of system interactions.
4. *Synchronization.* It is quite normal in a dynamic system that an event requires multiple resources. The resulting synchronization of resources can be captured by a transition with multiple input places. In Figure 9.5, t_4 is enabled only when each of its two input places receives a token. In general cases, the arrival of a token into each input place could be the result of a complex sequence of operations elsewhere in the rest of the Petri net model. Essentially, a transition of synchronization models a joining operation.
5. *Mutually exclusive.* Two processes are mutually exclusive if they cannot be performed at the same time due to constraints on the usage of shared resources. A real-world example is a robot shared by two machines for loading and unloading. Figure 9.6 shows this structure. The single token in place p dictates that at any moment, either the sequence of t_1 and t_2 is in execution or the sequence of t_3 and t_4 is in execution, but it will never be both.
6. *Priorities.* The classical Petri nets discussed so far have no mechanism to represent priorities. Such a modeling power can be achieved by introducing

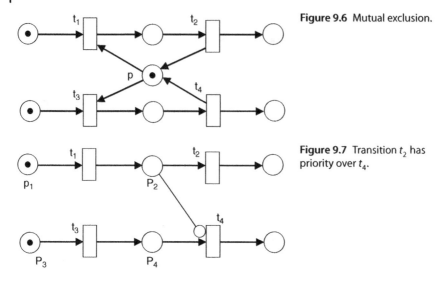

Figure 9.6 Mutual exclusion.

Figure 9.7 Transition t_2 has priority over t_4.

an *inhibitor arc*. The inhibitor arc connects an input place to a transition and is pictorially represented by an arc terminated with a small circle. The presence of an inhibitor arc connecting an input place to a transition changes the transition enabling conditions. In the presence of the inhibitor arc, a transition is regarded as enabled if each input place, connected to the transition by a normal arc (an arc terminated with an arrow), contains at least the number of tokens equal to the weight of the arc, and no tokens are present on each input place connected to the transition by the inhibitor arc. The transition firing rule is the same for normally connected places. The firing, however, does not change the marking in the places connected by the inhibitor arc. A Petri net with an inhibitor arc is shown in Figure 9.7. t_2 is enabled if p_2 contains a token, while t_4 is enabled if p_4 contains a token, and p_2 has no token. This gives priority to t_2 over t_4: In a marking in which both p_2 and p_4 have a token, t_4 won't be able to fire until t_2 is fired.

7. *Resource constraint.* Petri nets are well suited to model and analyze systems that are constrained by resources. For instance, Figure 9.8 depicts the two models of a writing–reading system. In the model (a), the transitions *write* and *send* can keep firing and injecting as many tokens (mails) as you want to the place *mail* that is connected to *send* and *receive*. Therefore, this model assumes an unbounded buffer or mailbox for mails between the sender and receiver. In the model (b), however, a place *mailbox* with three initial tokens is added to the Petri net, which limits to the consecutive firing times of *write* and *send* to only three. In fact, the mailbox is a resource in this system. The place *mailbox* here models the capacity of the mailbox. This example shows that resource constraints can be modeled very naturally in Petri nets.

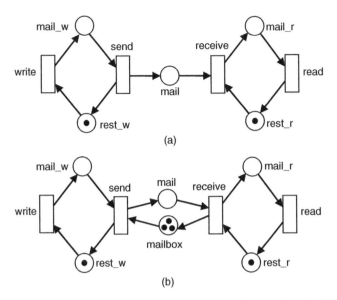

Figure 9.8 A writing–reading system. (a) The mailbox is unbounded. (b) The mailbox can hold up to three mails.

9.2 Petri Net Properties

As a mathematical tool, Petri nets possess a number of properties. These properties, when interpreted in the context of the modeled system, allow system designers to check if desired properties are in place, and meanwhile, undesired properties are avoided. Two types of properties can be distinguished: behavioral and structural. The behavioral properties depend on the initial state or marking of a Petri net. The structural properties, on the other hand, do not depend on the initial marking of a Petri net; they depend on its topology or structure instead.

9.2.1 Behavioral Properties

9.2.1.1 Reachability

An important issue in designing event-driven systems is whether a system can reach a specific state or exhibit a particular functional behavior. In general, the question is whether the system modeled with a Petri net exhibits all desirable properties as specified in the requirement specification and no undesirable ones.

In order to find out whether the modeled system can reach a specific state as a result of a required functional behavior, it is necessary to find such a transition firing sequence that would transform its Petri net model from the initial

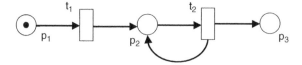

Figure 9.9 A Petri net with ω markings.

marking M_0 to a desired marking M_j, where M_j represents the specific state, and the firing sequence represents the required functional behavior. In general, a marking M_j is said to be *reachable* from a marking M_i if there exists a sequence of transition firings that transforms M_i to M_j. A marking M_j is said to be *immediately reachable* from M_i if firing an enabled transition in M_i results in M_j. The set of all markings reachable from a marking M is denoted by $R(M)$. We will explain how to obtain $R(M)$ later.

The reachability analysis of a bounded Petri net is conducted through the construction of reachability tree. Given a Petri net N, from its initial marking M_0, we can obtain as many "new" markings as the number of enabled transitions. From each new marking, we can again reach more markings. Repeating the procedure over and over results in a tree representation of the markings and firing transitions. Nodes represent markings generated from M_0 and its successors, and each arc represents the firing of a transition, which changes the Petri net from one marking to another.

9.2.1.2 ω Markings

The aforementioned tree representation, however, will grow infinitely large if the net is unbounded. To maintain the tree finite, we introduce a special symbol ω, which can be thought of as "infinity." It has the following properties:

- $\omega > n$
- $\omega + n = \omega$
- $\omega \geq \omega$

where n is any given integer.

For example, after t_1 is fired in the Petri net shown in Figure 9.9, the new marking is $(0, 1, 0)$. Now t_2 is enabled. Firing t_2 results in the marking $(0, 1, 1)$. Since in this marking, t_2 is still enabled, it can fire again, which results in $(0, 1, 2)$. Continuing to fire t_2, we will obtain $(0, 1, 3)$, $(0, 1, 4)$ ... Therefore, there are infinite number of markings with this Petri net. With the concept of ω markings, we use $(0, 1, \omega)$ to represent markings $(0, 1, n)$ for all $n \geq 1$.

The firing condition and firing rule for normal marking cases can be neatly extended to ω markings with the extended arithmetic rules. Basically, if a transition has an input place with ω tokens, then that place is considered to have sufficiently many tokens for the transition to fire, regardless of the arc weight. On the other hand, if a place contains ω tokens, then firing any transition that outputs tokens to the place will not change the number of tokens in the place.

9.2.1.3 Reachability Analysis Algorithm

Generally speaking, we don't know if a Petri net is bounded or not before we perform reachability analysis. However, we can construct a *coverability tree* if the net is unbounded or a reachability tree if the net is bounded according to the following general algorithm:

1. Label the initial marking M_0 as the root and tag it "new."
2. For every new marking M:
    - 2.1. If M is identical to a marking that already appeared in the tree, then tag M "old" and move on to another new marking.
    - 2.2. If no transitions are enabled at M, tag M "dead-end" and move on to another new marking.
    - 2.3. While there exist enabled transitions in M, do the following for each enabled transition t:
        - 2.3.1. Obtain the marking M' that results from firing t in M.
        - 2.3.2. On the path from the root to M, if there exists a marking M'' such that $M'(p) \geq M''(p)$ for each place p and $M' \neq M''$, that is, M'' is coverable, then replace $M'(p)$ by ω for each p such that $M'(p) > M''(p)$.
        - 2.3.3. Introduce M' as a node, draw an arc with label t from M to M', and tag M' "new."

If ω appears in a marking, then the Petri net is unbounded, and the tree is a coverability tree; otherwise, the net is bounded and the tree is a reachability tree. When all old nodes are merged with corresponding internal nodes, a reachability tree becomes a reachability graph, or a coverability tree becomes a coverability graph.

Example 9.3 *Reachability Tree and Reachability Graph*

Consider the Petri net shown in Figure 9.1 again. It has seven reachable markings:

$$M_0 = (2, 0, 0, 0),$$

$$M_1 = (0, 2, 1, 0),$$

$$M_2 = (0, 1, 1, 1),$$

$$M_3 = (0, 2, 0, 1),$$

$$M_4 = (0, 1, 0, 2),$$

$$M_5 = (0, 0, 1, 2),$$

$$M_6 = (0, 0, 0, 3).$$

The reachability tree of this Petri net is shown in Figure 9.10a, and the reachability graph is shown in Figure 9.10b.

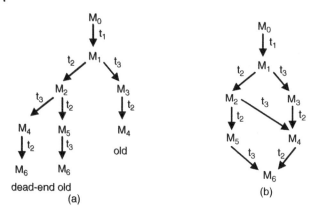

Figure 9.10 Reachability tree and reachability graph of the Petri net shown in Figure 9.1. (a) Reachability tree. (b) Reachability graph.

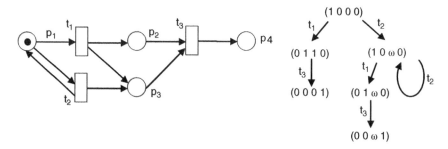

Figure 9.11 Coverability graph. (a) A Petri net. (b) Coverability graph.

Example 9.4 *Coverability Graph*

The Petri net shown in Figure 9.11 is an unbounded Petri net. Its initial marking $M_0 = (1, 0, 0, 0)$ enables t_1 and t_2. Firing t_1 results in a new marking $M_1 = (0, 1, 1, 0)$, which is a regular marking. Firing t_2 gives $(1, 0, 1, 0)$. Comparing it with its ancestor M_0, we know that $(1, 0, 1, 0) \geq M_0$. Therefore, we change the element that is increased to ω and obtain $M_2 = (1, 0, \omega, 1)$.

At M_1, t_3 is enabled. Firing t_3 results in a dead-end marking $M_3 = (0, 0, 0, 1)$.

At M_2, both t_1 and t_2 are enabled. Firing t_1 results in a new marking $(0, 1, \omega + 1, 0)$. Because $\omega + 1 = \omega$, we denote this marking as $M_4 = (0, 1, \omega, 0)$. Firing t_2 results in $(1, 0, \omega + 1, 0)$, which is equal to M_2.

At M_4, t_3 is enabled. Firing t_3 results $(0, 0, \omega - 1, 1)$. Because $\omega - 1 = \omega$, this is a dead-end marking $M_5 = (0, 0, \omega, 1)$.

Therefore, there are six markings with this Petri net, three being regular markings and three being ω markings:

$$M_0 = (1, \ 0, \ 0, \ 0),$$
$$M_1 = (0, \ 1, \ 1, \ 0),$$
$$M_2 = (1, \ 0, \ \omega, \ 1),$$
$$M_3 = (0, \ 0, \ 0, \ 1),$$
$$M_4 = (0, \ 1, \ \omega, \ 0),$$
$$M_5 = (0, \ 0, \ \omega, \ 1).$$

The coverability graph is shown in Figure 9.11b.

9.2.1.4 Boundedness and Safeness

In a Petri net, places are often used to represent information storage areas in communication and computer systems, product and tool storage areas in manufacturing systems, and so on. It is important to be able to determine whether the proposed control strategies prevent the overflows of these storage areas. The Petri net property that helps to identify the existence of overflows in the modeled system is *boundedness*.

A place p is said to be k-*bounded* if the number of tokens in p is always less than or equal to k (k is a nonnegative integer number) for every marking M reachable from the initial marking M_0, that is, $M \in R(M_0)$. It is *safe* if it is 1-bounded.

A Petri net $N = (P, T, I, O, M_0)$ is k-bounded (safe) if each place in P is k-bounded (safe). It is *unbounded* if k is infinitely large. For example, the Petri net shown in Figure 9.1 is 2-bounded, but the net of Figure 9.8a is unbounded.

9.2.1.5 Liveness

The concept of liveness is closely related to the *deadlock* situation, which has been situated extensively in the context of real-time embedded systems.

A Petri net model of a deadlock-free system must be *live*. This implies that for any reachable marking M, it is ultimately possible to fire any transition in the net by progressing through some firing sequence. This requirement, however, might be too strict to represent some real systems or scenarios that exhibit deadlock-free behavior. For instance, the initialization of a system can be modeled by a transition (or a set of transitions) that fires a finite number of times. After initialization, the system may exhibit a deadlock-free behavior, although the Petri net representing this system is no longer live as specified earlier. For this reason, different levels of liveness are defined. Denote the set of all possible firing sequences starting from M_0 by $L(M_0)$. A transition t in a Petri net is said to be:

L0-live (or dead) if there is no firing sequence in $L(M_0)$ in which t can fire.

L1-live (potentially firable) if t can be fired at least once in some firing sequence in $L(M_0)$.

L2-live if t can be fired at least k times in some firing sequence in $L(M_0)$ for any given positive integer k.

L3-live if t can be fired infinitely often in some firing sequence in $L(M_0)$.

L4-live (or live) if t is L1-live (potentially firable) in every marking in $R(M_0)$.

For example, all the three transitions in the net shown in Figure 9.1 are L1-live because both t_1 and t_3 can only fire once, while transition t_2 can fire twice. However, all transitions in the net shown in Figure 9.8a are L4-live, because they are all L1-live in every reachable marking.

9.2.2 Structural Properties

9.2.2.1 T-Invariants and S-Invariants

For a Petri net with m places and n transitions, we define its *incidence matrix* as follows:

$$A = \begin{bmatrix} a_{11} & \cdots & a_{1m} \\ \vdots & \ddots & \vdots \\ a_{n1} & \cdots & a_{nm} \end{bmatrix}$$

where

$$a_{ij} = O(t_i, \, p_j) - I(t_i, \, p_j)$$

A *T-invariant* is an integer solution x of the homogeneous equation

$$A^T x = 0$$

where A^T is the transpose of A. The nonzero entries in a T-invariant represent the firing counts of the corresponding transitions that belong to a firing sequence that first moves the Petri net away from M_0 and then brings it back to M_0.

A Petri net often has an infinite number of T-invariants. A T-invariant x is said to be *minimal* if there is no other T-invariant x' such that $x(t) \leq x'(t)$ for all t in T.

Example 9.5 *Incidence Matrix and T-Invariants*

The Petri net shown in Figure 9.12 has eight places and six transitions. Its incidence matrix is

$$A = \begin{bmatrix} -1 & 0 & -1 & 1 & 0 & 0 & 0 & 0 \\ 0 & -1 & -1 & 0 & 1 & 0 & 0 & 0 \\ 0 & 0 & 1 & -1 & 0 & 1 & 0 & 0 \\ 0 & 0 & 1 & 0 & -1 & 0 & 1 & 0 \\ 0 & 0 & 0 & 0 & 0 & -1 & -1 & 1 \\ 1 & 1 & 0 & 0 & 0 & 0 & 0 & -1 \end{bmatrix}$$

Figure 9.12 A Petri net with one minimal *T*-invariant.

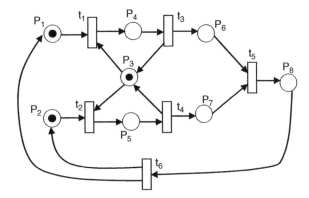

Its transpose is

$$A^T = \begin{bmatrix} -1 & 0 & 0 & 0 & 0 & 1 \\ 0 & -1 & 0 & 0 & 0 & 1 \\ -1 & -1 & 1 & 1 & 0 & 0 \\ 1 & 0 & -1 & 0 & 0 & 0 \\ 0 & 1 & 0 & -1 & 0 & 0 \\ 0 & 0 & 1 & 0 & -1 & 0 \\ 0 & 0 & 0 & 1 & -1 & 0 \\ 0 & 0 & 0 & 0 & 1 & -1 \end{bmatrix}$$

Let x be an 6 by 1 column vector $(x_1, x_2, x_3, x_4, x_5, x_6)^T$. It results from $A^T x = 0$ that

$$-x_1 + x_6 = 0$$
$$-x_2 + x_6 = 0$$
$$-x_1 - x_2 + x_3 + x_4 = 0$$
$$x_1 - x_3 = 0$$
$$x_2 - x_4 = 0$$
$$x_3 - x_5 = 0$$
$$x_4 - x_5 = 0$$
$$x_5 - x_6 = 0$$

Thus, we have $x_1 = x_2 = x_3 = x_4 = x_5 = x_6$. Because we are looking for a nonzero solution, we cannot assign 0 to x_1. Let $x_1 = k$, and k is any nonzero natural number, then we have $x = (k, k, k, k, k, k)^T$. Therefore, $x = (1, 1, 1, 1, 1, 1)^T$ is a solution, and it means that if we fire every transition in this Petri net once (in some order), it will return to the initial marking. By examining the Petri net, we can easily find that from the initial marking, the firing sequence $t_1 t_3 t_2 t_4 t_5 t_6$ or $t_2 t_4 t_1 t_3 t_5 t_6$ can each bring the net back to M_0.

Obviously, there are an infinite number of T-invariants for this Petri net. However, $(1, 1, 1, 1, 1, 1)^T$ is the only minimal invariant.

Example 9.6 *Petri Net with Multiple Minimal T-Invariants*

The Petri net shown in Figure 9.13 has two minimal T-invariants: $x1 = (1, 1, 1, 0)^T$ and $x2 = (1, 1, 0, 1)^T$. This can be easily verified by firing sequence $t_1 t_2 t_3$ and sequence $t_1 t_2 t_4$.

Not every Petri net has T-invariants. An example is the Petri net shown in Figure 9.1. For that net, the only vector that satisfies $A^T x = 0$ is $x = (0, 0, 0)^T$, which is not a valid T-invariant. Basically, the net does not show any repetitive behavior.

It is worth mentioning that a T-invariant only tells the firing count of each transition; it does not tell the transition firing sequence. Moreover, the existence of a T-invariant does not imply that it will actually be possible to fire the indicated transitions; the initial conditions of a net may prohibit it. Instead, a T-invariant indicates that if it is possible to fire the transitions in some order, the state of the net will return to its initial condition at the end of the sequence.

An *S-invariant* is an integer solution y of homogeneous equation

$$Ay = 0$$

The nonzero entries in an S-invariant represent weights associated with the corresponding places so that the weighted sum of tokens on these places is constant for all markings reachable from an initial marking.

An S-invariant y is said to be *minimal* if there is no other S-invariant y' such that $y(p) \geq y'(p)$ for all p in P.

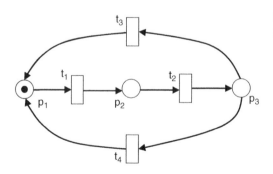

Figure 9.13 A Petri net with two minimal T-invariants.

Example 9.7   S-*Invariants*

The incidence matrix of the Petri net shown in Figure 9.13 is

$$A = \begin{bmatrix} -1 & 1 & 0 \\ 0 & -1 & 1 \\ 1 & 0 & -1 \\ 1 & 0 & -1 \end{bmatrix}$$

Let $y = (y_1, y_2, y_3)^{\mathrm{T}}$. Solving $Ay = 0$ gives

$$-y_1 + y_2 = 0$$
$$-y_2 + y_3 = 0$$
$$y_1 - y_3 = 0$$

Thus, $y = (1, 1, 1)^{\mathrm{T}}$ is a minimal S-invariant, which means that in all reachable markings, the sum of tokens in $p_1, p_2,$ and p_3 is a constant.

Invariants are important means for analyzing Petri nets since they allow for the net's structure to be investigated independently of any dynamic process.

9.2.2.2   Siphons and Traps

Let $\dot{}p = \{t | O(t, p) > 0\}$. $\dot{}p$ is called the *preset* of p. Let S be a subset of P, that is, $S \subseteq P$. Define

$$\dot{}S = \bigcup_{p \in S} \dot{}p$$

$\dot{}S$ is a set of transitions that output tokens to places in S. Similarly, let $p\dot{} = \{t | I(t, p) > 0\}$. $p\dot{}$ is called the *pos-set* of p. Define

$$S\dot{} = \bigcup_{p \in S} p\dot{}$$

$S\dot{}$ is a set of transitions that take tokens from places in S as input.

S is a *siphon* if $\dot{}S \subseteq S\dot{}$. Intuitively, if a transition is going to deposit a token to a place in a siphon S, the transition must also remove a token from S. A siphon is *minimal* if there is no siphon contained in it as a proper subset.

S is a *trap* if $S\dot{} \subseteq \dot{}S$. Intuitively, a trap S represents a set of places in which every transition consuming a token from S must also deposit a token back into S. A trap is *minimal* if there is no trap contained in it as a proper subset.

Siphons and traps are closely related to the reachability and potential deadlock of a Petri net. Once a siphon is emptied under a marking, it remains empty under subsequent markings. Once a trap is marked under a marking, it remains marked under subsequent markings. This is illustrated in the following example.

Example 9.8   *Siphons and Traps*

Let us examine three subsets of places in the Petri net shown in Figure 9.8b:

$$S_1 = \{mail\_w,\ rest\_w\}$$

$$S_2 = \{mail,\ mailbox\}$$

$$S_3 = \{mail\_r,\ rest\_r\}$$

Because $S_1 = \{write, send\} = S_1^{\cdot}$, S_1 is both a siphon and a trap. So are S_2 and S_3. Let us focus on S_2. For any reachable marking M, we always have $M(mail) + M(mailbox) = 3$. The sum of tokens in S_2 does not increase because S_2 is a siphon. Similarly, the sum of tokens in S_2 does not decrease because S_2 is a trap. Imagine if in the initial state, there is no token in *mailbox*, there will never be tokens added to S_2. In other words, markings that indicate that mails are sent by the writer are not reachable. This is, of course, undesired.

9.3   Timed Petri Nets

The need for including timing variables in the models of various types of dynamic systems is apparent since these systems are real-time in nature. In the real world, almost every event is time-related. A Petri net that contains time variables are called *timed Petri net*. The definition of a timed Petri net consists of three specifications:

- The topological structure
- The labeling of the structure
- Firing rules

The topological structure of a timed Petri net generally takes the form that is used in a conventional Petri net. The labeling of a timed Petri net consists of assigning numerical values to transitions, places, or arcs. The firing rules are defined differently depending on the way the Petri net is labeled with time variables. The firing rules defined for a timed Petri net control the process of moving the tokens around.

The aforementioned variations lead to several different types of timed Petri nets. Among them, deterministic timed Petri nets (DTPN) and time Petri nets (TPN), in which time variables are associated with transitions, are the two models widely used in real-time system modeling.

9.3.1   Deterministic Timed Petri Nets

A DTPN is a six-tuple (P, T, I, O, M_0, τ), where (P, T, I, O, M_0) is a Petri net, $\tau : T \to R^+$ is a function that associates transitions with deterministic time delays.

A transition t_i in a DTPN can fire at time τ if and only if:

- For any input place p of t_i, the number of tokens is greater than or equal to $I(t_i, p)$ in p continuously for the time interval $[\tau - \tau_i, \tau]$, where τ_i is the associated firing time of t_i;
- After the transition fires, each of its output places, p, will receive the number of tokens equal to $O(t_i, p)$ at time τ.

Let us use the DTPN shown in Figure 9.14 as an example to show the timeliness analysis. Transition firing times are as follows:

$$t_1 : 2 \quad t_2 : 1 \quad t_3 : 3 \quad t_4 : 3$$

At time 0, the initial marking (1, 0, 0, 0, 0, 0) has sufficient tokens to fire t_1, but the firing won't occur until time 2 because the firing time of t_1 is 2.

At time 2, t_1 fires, and the new marking is (0, 1, 1, 0, 0, 0). In this marking, both t_2 and t_3 have sufficient tokens to fire; however, only t_2 can fire in this marking, because its firing time is less than that of t_3.

At time 3, t_2 fires, and the new marking is (0, 0, 1, 1, 0, 0). In this marking, t_3 is the only transition that is enabled. Remember that p_3 is marked at time 2. Therefore, t_3 only needs to wait for 2 more time units to fire.

At time 5, t_3 fires, and the new marking is (0, 0, 0, 1, 1, 0). In this marking, t_4 has sufficient tokens to fire. Of course, it has to wait for 3 time units.

At time 8, t_4 fires. The new marking is (0, 0, 0, 0, 0, 1), and the Petri net hits a dead end. This also means that it takes 8 units of time for this Petri net to run from the initial marking to the final marking.

Example 9.9 *Scheduling of Periodic Tasks*

In this example, we show how to model the scheduling of periodic tasks. Consider two tasks T1 (5, 1) and T2 (10, 2) that are running on one processor. We first model the scheduling using the Petri net shown in Figure 9.15. The number in parentheses in each transition label indicates the transition firing time. The model shows two processes: the left-hand side half models the activities of T1, and the right-hand side half shows the activities of T2. They share a common processor that is modeled by the place *Processor*. The transition *T1 job arrives* and places *T1* and *T1 job* together model the job arrival process of the first periodic task. The transitions *T1 job scheduled* and *T2 job executed* and places

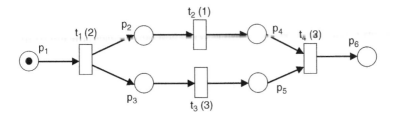

Figure 9.14 A deterministic timed Petri net.

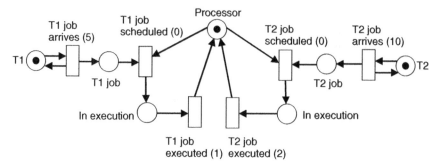

Figure 9.15 Timed Petri net model of two periodic tasks scheduled on one processor. First attempt.

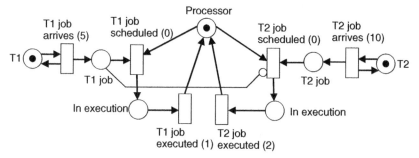

Figure 9.16 Timed Petri net model of two periodic tasks scheduled on one processor. Second attempt.

In execution and *Processor* together model the job execution process. The second task is modeled similarly.

We can see a problem with this model: it only specifies that the two periodic tasks share a single processor, but it does not reflect that jobs in T1 have a higher priority in getting executed on the processor than jobs in T2. For example, at time 10, there are jobs released from both T1 and T2, and thus, *T1 job* and *T2 job* each gets a token at the same time. According to this model, *T2 job scheduled* may be selected to fire, which violates the priority rule. To fix this problem, we introduce an inhibitor arc from *T1 job* to *T2 job scheduled*, which is shown in Figure 9.16. This ensures that when both tasks have a job released, the job from T1 will get scheduled and executed first.

Do we still miss anything in the updated model? Think about this case: a T1 job arrives while a T2 job is in execution. According to this model, the T1 job has to wait until the T2 job is finished before it can access the processor. In other words, this model does not allow high-priority jobs to *preempt* low-priority jobs.

To fix this problem, we have to extend the notation of timed Petri nets a little. First, we introduce "take-all" arcs that connect places to transitions. A take-all arc works this way: when the connected transition fires, it removes all tokens from the connected place, being it zero or more. Second, we introduce variable transition firing times – the firing time of a transition is determined by the value of a function. With these extensions, we build the final model of a periodic task scheduling problem as shown in Figure 9.17.

In the new model, we added a place *Count* that is connected as an output place of *T1 job scheduled* and input place of *T2 job scheduled*. The arc (*Count*, *T2 job scheduled*) is a take-all arc. As such, when transition *T2 job scheduled* is fired, all tokens in *Count* will be removed. Notice that whether there are tokens in *Count* does not matter. The purpose of this arc is to rather clean up the place when transition *T2 job scheduled* is fired. This way, the number of tokens in *Count* before transition *T2 job executed* fires indicates how many T1 jobs are released and executed before the T2 job is executed. That is why we set the firing time of transition *T2 job executed* to $2 + m$, where 2 is the duration of T2 jobs, while m is the maximum number of tokens in *Count* while the place *In execution* is marked. If the execution time of *T1 job executed* is τ and $\tau \neq 0$, then the execution time of *T2 job executed* should be set to $2 + m\tau$.

We also split the single place *Processor* in the previous two models to two, each with a token. The consequence is that T1 jobs are scheduled and executed as if task T2 didn't exist. T2 jobs are also running independently, except (1) when T1 and T2 release jobs at the same time, T1 job is scheduled first; (2) when a T2 job is scheduled, its completion time may be delayed due to the release and execution of T1 jobs. We know that this is consistent with what are discussed in Chapter 4.

9.3.1.1 Performance Evaluation Based on DTPNs

An important application of DTPNs is to calculate the cycle time of a class of systems in which job arrival times and job service times are known in advance of

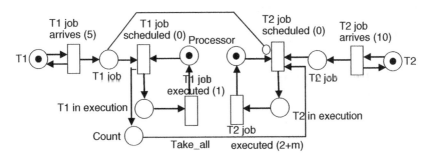

Figure 9.17 Timed Petri net model of two periodic tasks scheduled on one processor. m: maximum number of tokens in place *Count* while place *T2 in execution* is marked.

the analysis. Before we proceed, let us introduce some concepts first. In a Petri net, a sequence of places and transitions, $p_1 t_1 p_2 t_2 \ldots p_k$, is a *directed path* from p_1 to p_k if transition t_i is both an output transition of p_i and an input transition of p_{i+1} for $1 \leq i \leq k - 1$. If p_1 and p_k are the same place, but all other nodes in the directed path are different, then the path is a *directed circuit*. If in a Petri net, every place has exactly one input transition and one output transition then the Petri net is a *decision-free* Petri net or a *marked graph*.

Decision-free Petri nets have two unique properties. First, they are *strongly connected*, that is, there is a directed path between any two nodes in such a Petri net. Second, the total number of tokens in a directed circuit remains the same after any firing sequence. This is because whenever a transition in a circuit fires, it removes one and only one token from its input place in the circuit and adds one and only one token to its output place in the circuit.

Let $S_i(n_i)$ be the time at which a transition t_i initiates its n_i -th firing. Then, the *cycle time* C_i of the transition t_i is defined as

$$C_i = \lim_{n_i \to \infty} \frac{S_i(n_i)}{n_i}.$$

It has been proved that *all* transitions in a decision-free Petri net have the same cycle time C. Consider a decision-free Petri net with q directed circuits. For a circuit L_k, denote the sum of the firing times of all transitions in the circuit by T_k and the total number of tokens in all places of the circuit by N_k, that is,

$$T_k = \sum_{t_i \in L_k} \tau_i$$

$$N_k = \sum_{p_i \in L_k} M(p_i)$$

Both T_k and N_k are constants. N_k can be counted at the initial marking. Obviously, the number of transitions that are enabled simultaneously in L_k is less than or equal to N_k. On the other hand, the processing time required by circuit L_k per cycle, which is T_k, is less than or equal to the maximum processing power of the circuit per cycle time, which is CN_k. Therefore, we have

$$T_k \leq CN_k,$$

or

$$C \geq \frac{T_k}{N_k}$$

The bottleneck circuit in the decision-free Petri net is the one that satisfies $T_k = CN_k$. Therefore, the minimum cycle time C is given by

$$C = \max \left\{ \frac{T_k}{N_k} : k = 1, 2, \ldots, q \right\},$$

which corresponds to the best performance of the system modeled by the Petri net.

Example 9.10   *A Communication Protocol*

Consider the communication protocol between two processes: one indicated as the sender and the other as the receiver. The sender sends messages to a buffer, while the receiver picks up messages from the buffer. When it receives a message, the receiver sends an ACK back to the sender. After receiving the ACK from the receiver, the sender begins processing and sending a new message. Suppose that the sender takes 1 unit of time to send a message to the buffer, 1 unit of time to receive the ACK, and 3 units of time to process a new message. The receiver takes 1 time unit to receive the messages from the buffer, 1 unit of time to send an ACK back to the buffer, and 4 time units to process a received message. The DTPN model of this protocol is shown in Figure 9.18. The legends of places and transitions and timing properties are listed in Table 9.1.

There are three circuits in the model:
Circuit L_1: $p_1 t_1 p_3 t_5 p_8 t_6 p_1$. Its cycle time is

$$C_{L1} = \frac{T_1}{N_1} = \frac{1 + 1 + 3}{1} = 5,$$

Circuit L_2: $p_1 t_1 p_2 t_2 p_4 t_3 p_7 t_5 p_8 t_6 p_1$. Its cycle time is

$$C_{L2} = \frac{T_2}{N_2} = \frac{1 + 1 + 1 + 1 + 3}{1} = 7,$$

Circuit L_3: $p_5 t_2 p_4 t_3 p_6 t_4 p_5$. Its cycle time is

$$C_{L3} = \frac{T_3}{N_3} = \frac{1 + 1 + 4}{1} = 6.$$

After enumerating all circuits in the net, we know that the minimum cycle time of the protocol between the two processes is 7 time units.

Figure 9.18 Petri net model of a communication protocol.

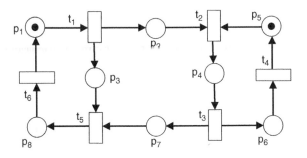

Table 9.1 Legend for Figure 9.18.

Place	Description
P_1	The sender ready
P_2	Message in the buffer
P_3	The sender waiting for ACK
P_4	Message received
P_5	The receiver ready
P_6	ACK sent
P_7	ACK in the buffer
P_8	ACK received

Transition	Description	Time delay
t_1	The sender sends a message to the buffer	1
t_2	The receiver receives the messages from the buffer	1
t_3	The receiver sends an ACK back to the buffer	1
t_4	The receiver processes the message	4
t_5	The sender receives the ACK	1
t_6	The sender processes a new message	3

9.3.2 Time Petri Nets

TPNs are first introduced by Merlin and Farber in 1976. In a TPN, two time values are defined for each transition, α^s and β^s, where α^s is the minimum time the transition must wait for after it is enabled and before it is fired, and β^s is the maximum time the transition can wait for before firing if it is still enabled. Times α^s and β^s, for a transition t, are relative to the moment at which t is enabled. Assuming that t has been enabled at time τ, then t, even if it is continuously enabled, cannot fire before time $\tau + \alpha^s$ and must fire before or at time $\tau + \beta^s$, unless it is disabled before its firing by the firing of another transition.

TPNs have been proven very convenient for constraints that are difficult to express except in terms of firing durations. Using TPNs, action synchronization is represented in terms of a set of pre- and postconditions associated with each individual action of the system under discussion, and timing constraints are expressed in terms of minimum and maximum times between the enabling and the execution of each action. This facilitates model specification by permitting a compact representation of the state space and an explicit modeling of concurrency and parallelism. Therefore, TPNs have gained application to the modeling and verification of real-time concurrent systems.

Mathematically, a TPN is a six-tuple (P, T, I, O, M_0, SI) where:

- (P, T, I, O, M_0) is a Petri net;
- SI is a mapping called *static interval*

$$SI : T \to Q^* \times (Q^* \cup \infty),$$

where Q^* is the set of positive rational numbers.

To analyze a TPN, it is necessary to differentiate static intervals and dynamic intervals associated with transitions. For each transition t, its static interval (or static firing interval) is defined as

$$SI(t) = (\alpha^s, \beta^s)$$

where α^s and β^s are rational numbers such that

$$0 \leq \alpha^s < +\infty,$$

$$0 \leq \beta^s < +\infty,$$

$$\alpha^s \leq \beta^s \ if \ \beta^s \neq \infty, \ \ or$$

$$\alpha^s < \beta^s \ if \ \beta^s = \infty.$$

The left bound α^s is called the *static earliest firing time* (SEFT for short), and the right bound β^s is called the *static latest firing time* (SLFT for short).

In the general case, in a marking other than the initial marking, the *dynamic firing interval* of a transition in the firing domain will be different from its static firing interval. The lower bound of the dynamic interval is called *dynamic earliest firing time* (EFT), and the upper bound is called the *dynamic latest firing time* (LFT), written as α and β, respectively.

For a transition t, α^s, β^s, α, and β are relative to the moment when t is enabled in a state. If t is enabled at an absolute time τ_{abs}, then t cannot fire before time $\tau_{abs} + \alpha^s$ or $\tau_{abs} + \alpha$ and must fire before or at the latest at time $\tau_{abs} + \beta^s$ or $\tau_{abs} + \beta$. t may be disabled by firing another transition t_m, which leads to a new marking at a different absolute time τ'_{abs}.

In a TPN model, firing a transition takes no time to complete: firing a transition at time τ leads to a new state at the same time τ. Furthermore, if a pair (α^s, β^s) is not defined for a transition, then it is implicitly assumed that the corresponding transition is a classical Petri net transition, and by default, it is associated with the pair $(\alpha^s = 0, \ \beta^s = +\infty)$.

9.3.2.1 States in a Time Petri Net

A general form for a state S of a TPN can be defined as a pair $S = (M, I)$. M is a marking, as what is defined in a regular Petri net. I is a set of inequalities; each inequality describes the upper bound and lower bound of the firing time of an enabled transition. The number of entries of I is given in the number of the enabled transitions in the marking M. Because different markings may have different numbers of enabled transitions, the number of entries in I varies from state to state.

For example, Figure 9.19 shows a simple TPN, in which

$$SI(t_1) = [4, 6],$$

$$SI(t_2) = [3, 5],$$

$$SI(t_3) = [2, 3].$$

For t_1, SEFT $= 4$ and SLFT $= 6$. For t_2, SEFT $= 3$ and SLFT $= 5$. For t_3, SEFT $= 2$ and SLFT $= 3$. The initial marking, $M_0 = (1\ 1\ 0\ 0\ 0)$, defines the initial state S_0 of the TPN, in which t_1 and t_2 are enabled. I_0 is given by

$$I_0 = \{4 \leq \theta(t_1) \leq 6,$$

$$3 \leq \theta(t_2) \leq 5\}.$$

9.3.2.2 Enabling and Firing Conditions of Transitions

A transition t is enabled in a marking M if for each $p \in P$, $M(p) \geq I(t, p)$. This rule is the same as for traditional Petri nets.

Assume that a transition t is enabled at time τ and remains continuously enabled in a state $S = (M, I)$. It is *firable* at time $\tau + \theta$ if and only if the relative firing time θ, relative to the absolute enabling time τ, is not smaller than the EFT of t, denoted by $EFT(t)$, and not greater than the smallest of the LFTs of all the transitions enabled in M, that is,

$$EFT(t) \leq \theta \leq \min\{LFT(t_k)|t_k \in E(M)\}.$$

This condition simply reflects the rule that an enabled transition can fire no earlier than its EFT and must fire no later than its LFT unless another transition fires and changes the marking M and thus the state S.

For example, in the initial state of the TPN shown in Figure 9.19,

$$\min\{LFT(t_k)|t_k \in E(M_0)\} = \min\{6, 5\} = 5.$$

Therefore, t_1 can only fire in the interval of $[4, 5]$, and t_2 can only fire in the interval of $[3, 5]$.

Delay θ is not a global time; it can be viewed as provided by a "virtual clock," local to the transition, which must have the same time unit (e.g., in terms of seconds) as τ.

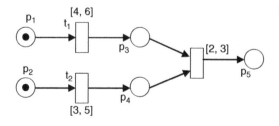

Figure 9.19 A TPN.

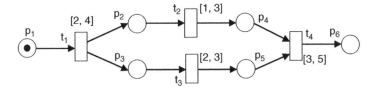

Figure 9.20 A time Petri net.

9.3.2.3 Firing Rules

Given a state $S = (M, I)$. Some transitions in $E(M)$ may never be able to fire due to the firing constraints of transitions (EFTs and LFTs). For example, if we change $SI(t_2)$ to [2, 3] in the TPN shown in Figure 9.19, then t_1 won't be able to fire in the initial state, because $\min\{6, 3\} = 3$, earlier than the $EFT(t_1)$.

Suppose that a transition t is firable at time $\tau + \theta$ in the $S = (M, I)$, and firing t in S results in an $S' = (M', I')$. The new state is calculated as follows:

1. M' is calculated with the same rule as for regular Petri nets:

$$(\forall p)M'(p) = M(p) - I(t_i, p) + O(t_i, p)$$

2. I' is computed in three steps:
    (a) Remove the entries that are related to the transitions that become disabled when t is fired, including t, from the expression of I.
    (b) Shift all remaining firing intervals in I by the value θ toward the origin of time, and truncate them, when necessary, to nonnegative values.
    (c) Introduce the domain new entries, each corresponding to the static interval of a newly enabled transition.

It should be easy to understand 2(a) and 2(c): when state changes, some originally enabled transitions may become disabled, and meanwhile, some originally disabled transitions may become enabled. Step 2(b) takes care of these transitions that are enabled in both S and S'. At S', their dynamic intervals are different from they were in S, because from S to S', a time of θ has elapsed, so their lower bounds and upper bounds need to be deducted by θ. If a lower bound becomes negative, change it to 0. Consider again the TPN shown in Figure 9.17, if t_1 fires at time 4.5 ($\theta = 4.5$), then in the new state, t_2 is still enabled, but its dynamic firing interval becomes [0, 0.5].

Example 9.11 *Time Petri Net Analysis*

Consider the TPN shown in Figure 9.20. Its initial state (M_0, I_0) is

$$M_0 = (1, 0, 0, 0, 0, 0),$$

$$I_0 = \{2 \le \theta(t_1) \le 4\}.$$

t_1 is the only enabled transition. After t_1 fires, the new state (M_1, I_1) is

$$M_1 = (0, 1, 1, 0, 0, 0),$$

$$I_1 = \{1 \le \theta(t_2) \le 3,$$
$$2 \le \theta(t_3) \le 3\}.$$

I_1 has two new entries because both t_2 and t_3 are enabled after the firing of t_1. These two transitions are in concurrent; firing one won't disable the other. If we fire t_2 at some time, say q, between 1 and 3, it will result in a new state (M_2, I_2), with

$$M_2 = (0, 0, 1, 1, 0, 0),$$
$$I_2 = \{\max\{0, 2 - q\} \le \theta(t_3) \le 3 - q\}$$

Notice that here we shifted the firing time interval of t_3 by q, according to the firing rule 2(b).

At state (M_2, I_2), t_3 is the only enabled transition. Firing t_3 results in state (M_3, I_3), in which

$$M_3 = (0, 0, 0, 1, 1, 0),$$
$$I_3 = \{3 \le \theta(t_4) \le 5\}.$$

t_4 is a newly enabled transition; it is added to I_3 with its static firing interval. Firing t_4 results in state (M_4, I_4), in which

$$M_4 = (0, 0, 0, 0, 0, 1),$$
$$I_4 = \varnothing.$$

TPNs offer the capability of specifying the lower bound and upper bound of event times. This is a very useful feature for real-time system specification. For example, it is common that there is jitter in task instance release times and execution times, even if the tasks are designed as periodic. Using an interval to specify a task's period or execution time, instead of a constant, will be more accurate and allow the designers to evaluate the effect of jitter on task scheduling and system performance.

Exercises

1 Construct the reachability trees of the three Petri nets shown in Figure 9.21.

2 Construct the coverability graphs of the three Petri nets shown in Figure 9.22.

3 Use a Petri net to model a manufacturing system with a single machine and buffer. Events with the system include the following:

Figure 9.21 Three Petri nets with regular markings only.

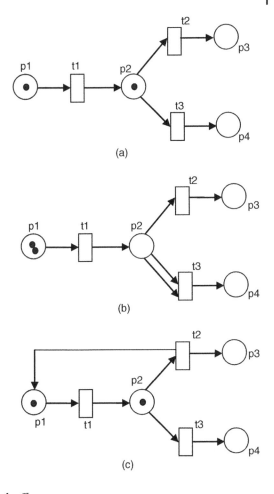

(a)

(b)

(c)

- A part arrives into the buffer.
- The machine starts processing.
- The machine ends processing.
- During processing, the machine may fail.
- If the machine fails, it will be repaired.
- After the machine is repaired, the processing continues.

Assume that the buffer can hold up to three parts. When the machine starts processing a part, one buffer slot is freed up for a new part.

4 Consider the classic dining philosophers problem. As illustrated in Figure 9.23, five silent philosophers sit at a round table with bowls of spaghetti. There are also forks on the table, each between two adjacent philosophers. However, the spaghetti is of a particularly slippery type,

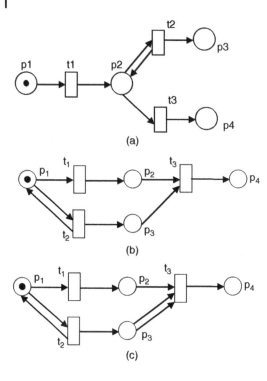

Figure 9.22 Three Petri nets with ω markings.

and a philosopher can only eat spaghetti when he has both left and right forks. The philosophers have agreed on the following protocols to grab the forks: Initially, they think about philosophy. When one gets hungry, he takes the fork on his left-hand side first, and then takes the one on his right-hand side, and then starts eating. He returns the two forks simultaneously to the table when he finishes eating and gets back to think about philosophy again. Of course, each fork can only be held by one philosopher at a time, and so when a philosopher tries to grab a fork, it may or may not be available. Model the behavior of the philosophers by a Petri net.

5 Consider a cruise control (CC) system in an auto. The CC controller has four buttons:

CC, Set, Cancel, and *Resume.*

To start any cruise control functions, the *CC* button has to be pressed, which brings the cruise control system from the *Off* state to the *Armed* state.

Figure 9.23 Five dining philosophers.

- At the *Armed* state, if the *Set* button is pressed, the system enters *Speed Control* state; if the *CC* button is pressed, the system goes back to the *Off* state.
- At the *Speed Control* state, if the *Cancel* button is pressed or the brake pedal is applied, the system changes to the *Cancelled* state; if the gas pedal is applied, then the system changes to the *Override* state.
- At the *Cancelled* state, if the *Resume* button is pressed, the system goes back to the *Speed Control* state; if the *CC* button is pressed, it goes back to the *Off* state.
- At the *Override* state, if the *Resume* button is pressed, the system goes back to the *Speed Control* state; if the *CC* button is pressed, it goes back to the *Off* state; if the *Cancel* button is pressed, it switches to the *Cancelled* state.

Model the behavior of the cruise controller with a Petri net.

6   Consider the classic ferryman puzzle. A ferryman has to bring a goat, a wolf, and a cabbage from the left bank to the right bank of a river. The ferryman can cross the river either alone or with exactly one of these three passengers. At any time, either the ferryman should be on the same bank as the goat, or the goat should be alone on a bank. Otherwise, the goat will eat the cabbage or the wolf will eat the goat. In Figure 9.24, we use places ML, WL, GL, and CL to model the ferryman, wolf, goat, and cabbage on the left bank, respectively. Similarly, we use the places MR, WR, GR, and CR to model the ferryman, wolf, goat, and cabbage on the right bank, respectively. Tokens in MR, WR, GR, and CR indicate that initially the four agents are all on the left bank. Transition MLR models the event that the ferryman travels alone to the right bank. Transition MGLR models the event that the ferryman travels to the right bank with the goat.

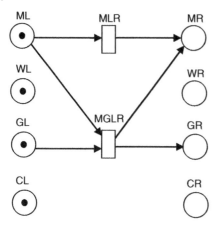

ML   MLR   MR

WL   WR

GL   MGLR   GR

CL   CR

Figure 9.24 Ferryman crosses river (incomplete).

(1) Model the event that the ferryman travels from the right bank to the left bank with the goat.

(2) Model the event that the goat eats the cabbage on the left bank. Be sure to model all preconditions and postconditions for the event.

(3) Model the event that the wolf eats the goat on the right bank. Be sure to model all preconditions and postconditions for the event.

(4) Find a sequence of transitions that enables the ferryman to bring all the passengers safely to the right bank.

7 Find out the liveness of each transition in the four Petri nets shown in Figure 9.25.

8 Find minimal T-invariants and S-invariants in the Petri net shown in Figure 9.26. Notice that $I(t_2, p_5) = 2$ and $O(t_3, p_5) = 3$.

9 In the Petri net shown in Figure 9.27, let

$$S_1 = \{p_1, p_2, p_3\}$$
$$S_2 = \{p_1, p_2, p_4\}$$
$$S_3 = \{p_1, p_2, p_3, p_4\}$$
$$S_4 = \{p_2, p_3\}$$
$$S_5 = \{p_2, p_3, p_4\}$$

Is each of the sets a siphon and/or a trap?

10 Consider again the three Petri nets shown in Figure 9.21. Assume that the firing times of t_1, t_2, and t_3 are 2, 4, and 3, respectively. Perform timeliness analysis for each Petri net.

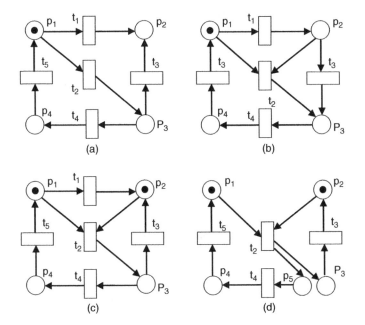

Figure 9.25 Four Petri nets for Problem 7.

Figure 9.26 Petri net for Problem 8.

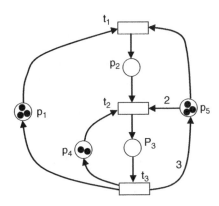

11 Consider three periodic tasks: T1 (3, 1), T2 (5, 1), and T3 (8, 1). They are scheduled on a single processor.
   (a) Assume that they are nonpreemptive. Draw the rate-monotonic scheduling model with a DTPN.
   (b) Assume that they are preemptive. Draw the rate-monotonic scheduling model with a DTPN.

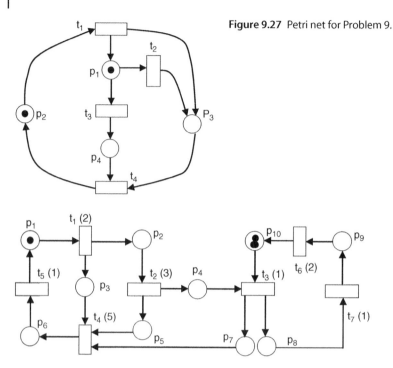

Figure 9.27 Petri net for Problem 9.

Figure 9.28 A deterministic timed Petri net.

12 Consider the decision-free DTPN shown in Figure 9.28.
  (a) List all directed circuits.
  (b) Calculate the minimum cycle time of this DTPN. Which circuit is the bottleneck circuit in terms of the processing rate of the described system?
  (c) Add a token to p_4. Recalculate the minimum cycle time.

Suggestions for Reading

The concept of Petri nets were introduced by Dr Carl Adam Petri in his dissertation in 1992 [1]. The first book on Petri nets was published in 1981, authored by J. L. Peterson [2]. In 1989, Dr T. Murata published a comprehensive survey paper on Petri nets in IEEE Proceedings [3], which has been cited intensively. DTPNs were first introduced in Ref. [4]. TPNs were proposed in Ref. [5]. The state-class-based solution to TPNs was discussed in Ref. [6]. A solution based on clock-stamped state class was presented in Ref. [7]. High-level Petri nets and colored Petri nets were introduced in Refs [8–10]. Exponentially distributed stochastic Petri nets were introduced in Refs [11, 12], and their applications can be found in Refs [13, 14]. Timing constraint with Petri nets and their application to schedulability analysis of real-time system specifications were discussed in Ref. [15]. All different types of timed Petri net models were presented in Ref. [16].

References

1 Petri, C.A. (1962) Kommunikation mit Automaten. Technical Report RADC-TR-65-377, Rome Air Dev. Center, New York.

2 Peterson, J.L. (1981) *Petri Net Theory and the Modeling of Systems*. N.J.: Prentice-Hall.

3 Murata, T. (1989) Petri nets: properties, analysis and applications. *Proceedings of the IEEE*, **77**(4): 541–580.

4 Ramamoorthy, C. and Ho, G. (1980) Performance evaluation of asynchronous concurrent systems using Petri nets. *IEEE Transaction on Software Engineering*, **SE-6** (5), 440–449.

5 Merlin, P. and Farber, D. (1976) Recoverability of communication protocols - implication of a theoretical study. *IEEE Transactions on Communications*. **24**(9):1036–1043.

6 Berthomieu, B. and Diaz, M. (1991) Modeling and verification of time dependent systems using time Petri nets. *IEEE Transactions on Software Engineering*, **17** (3), 259–273.

7 Wang, J., Deng, Y., and Xu, G. (2000) Reachability analysis of real-time systems using time Petri nets. *IEEE Transactions on Systems, Man, and Cybernetics*, **B30** (5), 725–736.

8 Genrich, J.H. and Lautenbach, K. (1981) System modeling with high-level Petri nets. *Theoretical Computer Science*, **13**, 109–136.

9 Jensen, K. (1981) Colored Petri nets and the invariant-method. *Theoretical Computer Science*, **14**, 317–336.

10 Jensen, K. (1997) Coloured Petri Nets: Basic Concepts, Analysis Methods and Practical Use *(3 volumes)*, Springer-Verlag, London.

11 Molloy, M. (1981) On the integration of delay and throughput measures in distributed processing models. Ph.D. Thesis, UCLA.

12 Natkin, S. (1980) Les Reseaux de Petri Stochastiques et Leur Application a l'evaluation des Systemes Informatiques. These de Docteur Ingegneur, Cnam, Paris, France.

13 Ajmone Marsan, M. (1990) Stochastic Petri nets: an elementary introduction. *Advances in Petri Nets*, LNCS, **424**, 1–29.

14 Molloy, M. (1982) Performance analysis using stochastic Petri nets. *IEEE Transactions on Computers*, **31** (9), 913–917.

15 Tsai, J., Yang, S., and Chang, Y. (1995) Timing constraint Petri nets and their application to schedulability analysis of real-time system specifications. *IEEE Transactions on Software Engineering*, **21**(1): 32–49.

16 Wang, J. (1998) *Timed Petri Nets: Theory and Application*, Kluwer Academic Publishers, Boston.

10

Model Checking

Model checking is an automatic verification technique for finite-state concurrent and reactive systems. It is developed to verify if assertions on a system are true or not. This is compared at program testing or simulation, which is to find out if there are bugs in a system. The aim of this chapter is to introduce the model checking technique. Because model checking is based on temporal logic, linear temporal logic (LTL), computation tree logic (CTL), and real-time computation tree logic (RTCTL) are introduced. A model checking tool, NuSMV, and its associated system description language are also presented.

10.1   Introduction to Model Checking

Chapter 1 discussed the importance of the reliability and correctness of real-time embedded systems. The two most common approaches to ensure software correctness are testing and simulation. Software testing involves the execution of a software component or system component to evaluate one or more properties of interest. This approach is very useful in practice, although it is clearly not possible to use it in highly critical systems if the testing data could cause damages in case of errors before real deployment. Simulation is based on the process of modeling a real system with a set of mathematical formulas. It is, essentially, a program that allows the user to observe an operation through simulation without actually performing that operation. Simulation does not work directly on the real system, which is a big advantage over testing.

Both testing and simulation are widely applied in industrial applications. However, program testing or simulation can only show the presence of errors but never their absence. It is not possible, in general, to simulate or test all the possible scenarios or behaviors of a given system. In other words, those techniques are not exhaustive due to the high number of possible cases to be taken into account, and the failure cases may be among those not tested or simulated.

Real-Time Embedded Systems, First Edition. Jiacun Wang.
© 2017 John Wiley & Sons, Inc. Published 2017 by John Wiley & Sons, Inc.

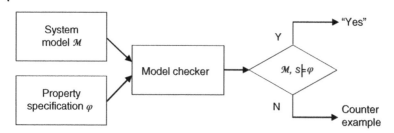

Figure 10.1 Model checking.

Model checking is an automatic verification technique for finite-state concurrent systems. It was originated independently by the pioneering work by E. M. Clarke and E. A. Emerson and by J. P. Queille and J. Sifakis in the early 1980s. In model checking, digital circuits or software designs under study are modeled as state transition systems, and desired properties are specified with temporal logic formulas. Verification is performed to find out whether the finite-state model meets the specifications. Verification is carried out by running the model checking tools, which are also called *model checkers*.

Model checking technique is based on solid mathematics, logics, and computer science. It is a formal verification method and can be stated as follows: given a desired property, expressed as a temporal logic formula φ, and a structure M, decide if $M, s \models \varphi$, that is, if φ is satisfied by M from a given state s. If it is, the model checking tool or model checker will simply output something as a "yes"; If it is not, the tool will print out a counterexample of execution in which the property is violated. This is illustrated in Figure 10.1.

10.2 Temporal Logic

Logic plays a fundamental role in computer science. It takes into account syntactically well-formed statement and studies whether they are semantically correct. Classic *propositional logic* deals with declarative sentences or *propositions*. Propositions can be formed by other propositions with the use of logical operators. An indivisible proposition is called an *atom*. Propositional logic is concerned with the study of the truth value of propositions and how their value depends on the truth value of their components. Propositions are evaluated in a single fixed state. Examples of propositions are

Earth is the center of the universe.
Five plus five is equal to ten and five minus five is equal to zero.
If it rains, then the ground is wet.

Temporal logic is any system of rules and symbolism for representing and reasoning about propositions qualified in terms of time. It describes the ordering of events in time without explicitly introducing time. In temporal logic, we can specify that

The elevator cannot move until the door is closed.
The program is in its critical section.
When a program enters a critical section, it will eventually leave the critical section.

Basically, a temporal logic statement or *formula* is not statically *true* or *false* in a model. A temporal logic model contains states, and a formula can be true in one state but false in another state. The set of states correspond to moments in time. How we navigate between these states depends on our particular view of time.

There are two models of time. One model thinks of time as a path of time points; for any two points, we can find that one is earlier than the other. Mathematically, we represent time as a structure $(T, <)$ such that $<$ is a precedence relation at T. Elements of T are time points. If a pair (s, t) belongs to $<$, we say that s is earlier than t or $s < t$. For a point t, the set $\{s \in T \mid t < s\}$ is called the *future* of t; the past of t is defined similarly. This model is called *linear-time model*, because all time points are linearly ordered. The linear-time model is illustrated in Figure 10.2a.

The second model is based on a *branching-time* structure. In this structure, a time point, say r, may have two or more future time points that are *not* related to each other. In other words, for any two of these future points, say s and t, we cannot say that s is the future of t, or vice versa. This also means that the future of r is not deterministic; or, r branches out to the future. The branching-time structure is a tree structure, rooted at the present moment, which is illustrated in Figure 10.2b.

We introduce two popular temporal logic models in this section. One is LTL that models time as sequence of states, and the other one is CTL that models time as a tree-like structure.

Figure 10.2 Modeling of time.
(a) Linear-time model.
(b) Branching-time model.

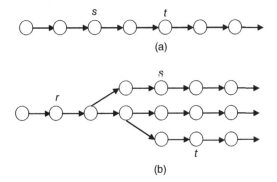

10.2.1 Linear Temporal Logic

LTL models time as a sequence of states, extending infinitely into the future. A sequence of states is also called a *computation path* or simply a *path*. Future is not determined, so there is more than one path, representing different possible futures.

10.2.1.1 Syntax of LTL

Recall that there are four propositional logical operators:

- ¬ for negation, or "not."
- ∧ for conjunction, or "and."
- ∨ for disjunction, or "or."
- → for implication, or "if–then."

Let φ and ψ be two propositional formulas. Table 10.1 shows the truth table for all these operators.

Temporal logic extends classical propositional logic with a set of temporal operators that navigate between states. The operators include the following:

- **X** for neXt state. **X** p is true if p is true in the next state.
- **F** for some Future state. **F** p is true if there is a reachable future state in which p is true.
- **G** for all future states (globally). **G** p is true if p is true in all future states.
- **U** for Until. p **U** q is true if p is true until q is true in a future state.
- **R** for Release. p **R** q is true if q is true until the first position in which p is true.

Assume a fixed set Σ of atomic propositional formulas and $p \in \Sigma$. An LTL formula is inductively defined in the Backus–Naur form as

$$\varphi ::= \top \,|\, \bot \,|\, p \,|\, (\neg\varphi) \,|\, (\varphi \wedge \psi) \,|\, (\varphi \vee \psi) \,|\, (\varphi \rightarrow \psi)$$
$$|\,(\mathbf{X}\varphi)\,|\,(\mathbf{F}\varphi)\,|\,(\mathbf{G}\varphi)\,|\,(\varphi\mathbf{U}\psi)\,|\,(\varphi\mathbf{R}\psi) \tag{10.1}$$

where $\top$ stands for *true* and $\bot$ stands for *false*. Equation (10.1) says that the two logic constants *true* and *false* and any atomic propositional formulas are LTL

Table 10.1 Truth table for propositional logic operators.

φ	ψ	$\neg\varphi$	$\varphi \wedge \psi$	$\varphi \vee \psi$	$\varphi \rightarrow \psi$
T	T	F	T	T	T
T	F	F	F	T	F
F	T	T	F	T	T
F	F	T	F	F	T

formulas; the negation of any LTL formula is an LTL formula; the conjunction of any two LTL formulas is an LTL formula, and so on.

The LTL definition also indicates that ¬, **X**, **F**, and **G** are unary operators, while all others are binary operators. Unary operators have the highest binding priority, followed by the temporal operators **X**, **F**, and **G**, then ∧ and ∨, and the last is →. Therefore, the formula

$$((\neg p) \wedge ((\mathbf{G}q) \vee (\neg q))) \rightarrow (p\mathbf{U}q)$$

is equal to

$$\neg p \wedge (\mathbf{G}q \vee \neg q) \rightarrow p\mathbf{U}q$$

An LTL formula is *syntactically correct* or *well-formed* if and only if it obeys the inductive construction rule given in the definition. For example, the following formulas are well-formed:

- $p \vee \neg (p \vee \mathbf{G}\,q) \rightarrow p$
- $\mathbf{F}\,(p \rightarrow \mathbf{X}\,r) \rightarrow q$
- $\mathbf{GF}\,(q\,\mathbf{R}\,r)$

The following formulas are not well-formed:

- $\mathbf{F}\,p \wedge \mathbf{G}\,q \rightarrow \mathbf{U}\,r$
- $\mathbf{G}\,(p \rightarrow q\,\mathbf{X}\,\mathrm{r})$
- $p\,\mathbf{U}\,(\wedge\,\mathbf{r})$

10.2.1.2 Parse Trees for LTL Formulas

A parse tree of an LTL formula is a nested list, where each branch is either a single atomic proposition or a formula composed of either two propositions and a binary operator or one proposition and a unary operator. Parse trees are helpful for the truth value evaluation of LTL formulas. The root of the parse tree of an LTL formula is the operator that should be evaluated last, while all leaf nodes of the tree are atomic propositions. In case the formula has only an atomic proposition, then the proposition is the root and only node. Parse trees are evaluated bottom-up.

Example 10.1   *Parse Tree*

Consider the formula $\neg p \wedge (\mathbf{G}\,q \vee \neg q) \rightarrow p\,\mathbf{U}\,q$. In this formula, the only → operator has the lowest binding priority and is evaluated last. Thus, it is the root of the parse tree. Its left branch is the subtree of $\neg p \wedge (\mathbf{G}\,q \vee \neg q)$, in which the only ∧ operator has the lowest binding priority, and thus, ∧ is the root of the subtree. The right branch of → is the subtree of $p\,\mathbf{U}\,q$, in which **U** is the only operator and p and q are atomic propositions, and thus, the root of the subtree is **U**. Continuing the parsing recursively until all operators are hit, we end up with a parse tree depicted in Figure 10.3.

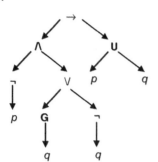

Figure 10.3 Parse tree of formula $\neg p \wedge (\mathbf{G} \, q \vee \neg q) \rightarrow p \, \mathbf{U} \, q$.

10.2.1.3 Semantics of LTL

LTL models time as an infinite sequence of states in which each point in time has a unique successor, based on a linear-time perspective. Again, we assume a fixed set $\sum$ of atomic propositions. For a set of states S, let L be a *labeling function* that maps S to the power set of $\sum$. The power set of $\sum$, denoted by $2^{\sum}$, is the set of all subsets of $\sum$. For example, if $\sum = \{p, q, r\}$, then

$$2^{\sum} = \{\varnothing, \, p, \, q, \, r, \, \{p, q\}, \, \{q, r\}, \, \{p, r\}, \, \{p, q, r\}\}$$

For each individual state s, $L(s)$ is a set of all atomic propositions that are evaluated to be true in the state.

To discuss the semantics of LTL, we consider a path

$$\pi = s_1 \rightarrow s_2 \rightarrow \cdots$$

and denote by π_i the path

$$\pi_i = s_i \rightarrow s_{i+1} \rightarrow \cdots$$

We now define the binary satisfaction relation, denoted by $\vDash$, for LTL formulas. The satisfaction is with respect to a computation path π. We always have

- $\pi \vDash \top$
- $\pi \not\vDash \bot$

For any single atomic proposition $p \in \sum$, $\pi \vDash p$ if only if (*iff* for short) $p \in L(s)$. For any LTL formula φ, we have $\pi \vDash \varphi$ iff φ evaluates to be true in s_1. Any composition of φ and ψ with propositional logic operators ($\neg$, $\wedge$, $\vee$, and $\rightarrow$) is evaluated in the first state s_1 of the path. Specifically,

- $\pi \vDash \neg\varphi$ iff $\pi \not\vDash \varphi$
- $\pi \vDash \varphi \wedge \psi$ iff $\pi \vDash \varphi$ and $\pi \vDash \psi$
- $\pi \vDash \varphi \vee \psi$ iff $\pi \vDash \varphi$ or $\pi \vDash \psi$
- $\pi \vDash \varphi \rightarrow \psi$ iff $\pi \vDash \psi$ as long as $\pi \vDash \varphi$

All LTL formulas that are composed of temporal operators should be evaluated across states. The operator $\mathbf{X}$ is formally defined as follows:

$$\pi \vDash \mathbf{X}\varphi \quad \text{iff} \quad \pi_2 \vDash \varphi$$

That is, iff φ is evaluated to be true in the next state s_2, then we have $\pi \models \mathbf{X}\varphi$.

The operator $\mathbf{G}$ is formally defined as follows:

$$\pi \models \mathbf{G}\varphi \quad \text{iff} \quad \pi_i \models \varphi \quad \text{for all } i \geq 1$$

That is, iff φ is evaluated to be true in every state along the path π, then we have $\pi \models \mathbf{G}\varphi$.

The operator $\mathbf{F}$ is formally defined as follows:

$$\pi \models \mathbf{F}\varphi \quad \text{iff} \quad \pi_i \models \varphi \quad \text{for some } i \geq 1$$

That is, iff φ is evaluated to be true in some state along the path π, then we have $\pi \models \mathbf{F}\varphi$.

The operator $\mathbf{U}$ is formally defined as follows:

$$\pi \models \varphi\mathbf{U}\psi \quad \text{iff} \quad \pi_i \models \psi \quad \text{for some } i \geq 1 \text{ and } \pi_j \models \varphi \quad \text{for all}$$
$$j = 1, 2, \ldots i - 1$$

That is, iff φ is evaluated to be true in every state along the path π until a state in which ψ is evaluated to be true, then we have $\pi \models \varphi\mathbf{U}\psi$. By the way, we can view $\mathbf{F}\varphi$ as an abbreviation of $\top\mathbf{U}\varphi$.

The operator $\mathbf{R}$ is formally defined as follows:

$$\pi \models \varphi\mathbf{R}\psi \quad \text{iff} \quad \pi_i \models \varphi \quad \text{for some } i \geq 1 \text{ and } \pi_j \models \psi \quad \text{for all } j = 1, 2, \ldots i$$

That is, iff ψ is evaluated to be true in every state along the path π until a state in which both φ and ψ are evaluated to be true, then we have $\pi \models \varphi\mathbf{R}\psi$.

Figure 10.4 illustrates the semantics of all these temporal operators. The first bubble in each path represents the first state of the path.

We use *transition systems* to model the systems that we want to verify. A transition system is defined as a Kripke structure $M = (S, I, R, L)$, where

- S is a set of states.
- I is a set of initial states $I \subseteq S$.

Figure 10.4 Illustration of semantics of LTL temporal operators.

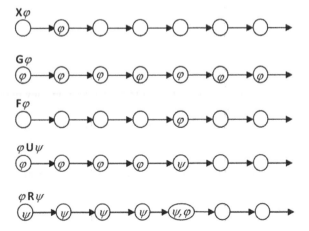

- $R \subseteq S \times S$ describes state transition relations. For each $s \in S$, there is s' such that $s \to s'$ and the relation is denoted as $(s, s') \in R$.
- L is a labeling function that maps S to the power set of $\sum$, where $\sum$ is the set of atomic propositions.

A transition system can be more intuitively illustrated as a directed graph, where each node is a state, state transitions are depicted by directed arrows, and the labeling of each state is marked on the node.

Example 10.2   *Transition Systems*
In the transition system shown in Figure 10.5, $\sum = \{p, q, r\}$, and the Kripke structure is

$$S = \{s_1, s_2, s_3, s_4\},$$
$$I = \{s_1\},$$
$$R = \{(s_1, s_2,), (s_1, s_3,), (s_2, s_1,), (s_2, s_4,), (s_3, s_4,), (s_4, s_3)\},$$
$$L(s_1) = \{p, q\},$$
$$L(s_2) = \{p, r\},$$
$$L(s_3) = \{q\},$$
$$L(s_4) = \{r\}.$$

Figure 10.5 Directed graph of a transition system.

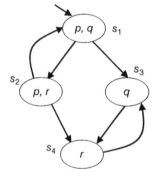

Unwinding the transition system results in an infinite tree of all possible computation paths, as shown in Figure 10.6.

As we can see from Example 10.2, a system may have many or even infinite number of computation paths. When we verify a system against an LTL formula from a state (typically an initial state), we check if the formula is satisfied by *all* paths from the state. We have $M, s \vDash \varphi$ if $\pi \vDash \varphi$ holds for every computation path π starting at s.

Note that sometimes we do list variables that are not true in a state in the graph of a transition system, just to improve the readability.

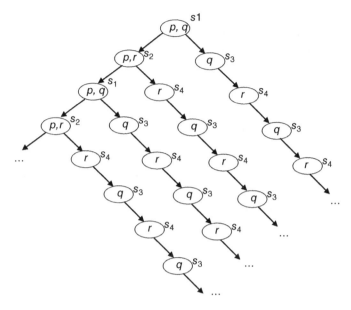

Figure 10.6 Computation paths of the system shown in Figure 10.5.

Example 10.3 *LTL Formula Verification*

We want to verify the model introduced in Example 10.2 against a few LTL specifications from the initial state s_1.

1. $\varphi = p \wedge q$.

   Because φ is essentially a propositional formula, $M, s_1 \models \varphi$ is evaluated only in s_1. Because $L(s_1) = \{p, q\}$, $M, s_1 \models p \wedge q$ holds.

2. $\varphi = p \wedge r$.

   Because s_1 does not satisfy r, $M, s_1 \models p \wedge r$ does not hold.

3. $\varphi = X q$.

   We check to see if along every path, the next state has q. Since s_2 does not have q, $M, s_1 \models X q$ does not hold.

4. $\varphi = F q$.

   We check to see if r is true in some state along every path. This is the case. Therefore, $M, s_1 \models F q$ holds.

5. $\varphi = G (q \vee r)$.

   In every state, either q is true or r is true. Therefore, $M, s_1 \models G (q \vee r)$ holds.

6. $\varphi = G(\neg p \rightarrow q \vee r)$.

   We check to see if for each state that satisfies $\neg p$, it also satisfies $q \vee r$. We know this is true, because all states that satisfy $\neg p$ have either q or r. Therefore, $M, s_1 \models G(\neg p \rightarrow q \vee r)$ holds.

7. $\varphi = \mathbf{FG}\,r$.

We check to see if there is a state in every path such that starting that state $\mathbf{G}\,r$ is true. Obviously, there is no single path that satisfies $\mathbf{G}\,r$ from some point along the path. Therefore, $M, s_1 \vDash \mathbf{FG}\,r$ does not hold.

8. $\varphi = \mathbf{GF}\,r$.

We check to see if $\mathbf{F}\,r$ is true in every state along every path, or in other words, if r is satisfied *infinitely often*. We know that this is true, because the state s_2 or s_4 appears in each path infinitely often. Therefore, $M, s_1 \vDash \mathbf{GF}\,r$ holds.

9. $\varphi = q\,\mathbf{U}\,r$.

For all paths starting with $s_1 \rightarrow s_2$, q is true in the first state and r is true in the second state, which satisfies $q\,\mathbf{U}\,r$. All other paths start with $s_1 \rightarrow s_3 \rightarrow s_4$ and also satisfy $q\,\mathbf{U}\,r$. Therefore, $M, s_1 \vDash q\,\mathbf{U}\,r$ holds.

10. $\varphi = (p \wedge q)\,\mathbf{U}\,(p \wedge r)$.

Note that $p \wedge r$ is true only in the states along the left-most path. In all other paths, it has never been true. Take the right-most path, for example. $p \wedge q$ is true only in the first state. In the second state, neither $p \wedge q$ nor $p \wedge r$ is true. Therefore, $M, s_1 \vDash (p \wedge q)\,\mathbf{U}\,(p \wedge r)$ does not hold.

11. $\varphi = \mathbf{F}\,q \rightarrow \mathbf{F}\,(p \wedge r)$.

This formula states that if any path starting from s_1 satisfies $\mathbf{F}\,q$, the path also satisfies $\mathbf{F}\,(p \wedge r)$. Check the right-most path. It satisfies $\mathbf{F}\,q$ but not $\mathbf{F}\,(p \wedge r)$. Therefore, $\varphi = \mathbf{F}\,q \rightarrow \mathbf{F}\,(p \wedge r)$ does not hold.

12. $\varphi = \mathbf{GF}\,q \rightarrow \mathbf{GF}\,r$.

This formula states that if any path starting from s_1 satisfies $\mathbf{GF}\,q$, the path also satisfies $\mathbf{GF}\,r$. The directed graph shows that every single path starting from s_1 satisfies $\mathbf{GF}\,q$; meanwhile, each of such paths also satisfies $\mathbf{GF}\,r$. Therefore, $M, s_1 \vDash \mathbf{GF}\,q \rightarrow \mathbf{GF}\,r$ holds.

10.2.1.4 Equivalencies of LTL Formulas

Two LTL formulas φ and ψ are said to be *semantically equivalent*, denoted by $\varphi \equiv \psi$, if for all models M and all states s in M,

$$M, s \vDash \varphi \text{ iff } M, s \vDash \psi.$$

Simply put, two formulas are equivalent if they are evaluated to the same truth value from any state of any computation path of any Kripke structure. A few equivalencies are as follows.

$$\mathbf{X}\,(\varphi \wedge \psi) \equiv \mathbf{X}\varphi \wedge \mathbf{X}\psi$$
$$\mathbf{X}\,(\varphi \vee \psi) \equiv \mathbf{X}\varphi \vee \mathbf{X}\psi$$
$$\mathbf{X}\,(\varphi\,\mathbf{U}\psi) \equiv \mathbf{X}\varphi\,\mathbf{U}\,\mathbf{X}\psi$$
$$\neg\,\mathbf{X}\varphi \equiv \mathbf{X}\neg\varphi$$

$$\mathbf{F}\,(\varphi\,\vee\,\psi)\;\equiv\;\mathbf{F}\varphi\,\vee\,\mathbf{F}\psi$$
$$\mathbf{G}\,(\varphi\,\wedge\,\psi)\;\equiv\;\mathbf{G}\varphi\,\wedge\,\mathbf{G}\psi$$
$$\neg\mathbf{F}\varphi\;\equiv\;\mathbf{G}\neg\varphi$$
$$\mathbf{F}\;\mathbf{F}\varphi\;\equiv\;\mathbf{F}\varphi$$
$$\mathbf{G}\;\mathbf{G}\varphi\;\equiv\;\mathbf{G}\varphi$$

10.2.1.5 System Property Specification

Let us use an elevator of five floors as an example to show how we can use LTL formulas to code the properties of real-world systems.

1. The elevator should not move if the door is open.

   ```
   G(¬door_closed → ¬(direction_up ∨ direction_down))
   ```

2. Whenever the door is open, it will eventually be closed.

   ```
   G(¬door_closed → F door_closed)
   ```

3. Similarly, whenever the door is closed, it will eventually be open.

   ```
   G(door_closed → F ¬door_closed)
   ```

4. The elevator can move upward only if the floor is not the highest.

   ```
   G(direction_up → ¬floor_5)
   ```

5. Similarly, the elevator can move downward only if the floor is not the lowest.

   ```
   G(direction_down → ¬floor_1)
   ```

6. When a floor button is pressed, the elevator will eventually stop at the floor. For example,

   ```
   G(button_3 → F floor_3)
   ```

7. When the elevator is traveling upward, it does not change its direction when it has passengers waiting to go to a higher floor. For example,

   ```
   G ((floor_1 ∨ floor_2 ∨ floor_3) ∧ direction_up ∧
        button_4
        → direction_up U floor_4)
   ```

8. Similarly, when the elevator is traveling downward, it does not change its direction when it has passengers waiting to go to a lower floor. For example,

   ```
   G ((floor_4 ∨ floor_5 ∨ floor_3) ∧ direction_down ∧
        button_2
        → direction_down U floor_2)
   ```

10.2.2   Computation Tree logic

LTL formulas are evaluated on paths. We say that a state of a system satisfies an LTL formula if *all paths* from the state satisfy it. If we want to specify that there exists a path that satisfies some property φ, we can verify if all paths satisfy $\neg\varphi$. A positive answer to the new problem is a negative answer to the original problem, and vice versa.

Branching-time logics solve this problem with the capability of explicitly quantifying over paths. CTL is a branching-time logic; It models time as a tree-like structure in which the future is not determined.

10.2.2.1   Syntax of CTL

CTL formulas are composed of propositional operators and CTL temporal operators. As shown in Figure 10.7, each CTL temporal operator is a pair of symbols. The first one is a *path quantifier*. It can be either **A** ("for All paths") or **E** ("there Exists a path"). The second one is a temporal operator and can be **X** ("neXt state"), **F** ("in a Future state"), **G** ("Globally in the future"), or **U** ("Until"), exactly as defined in LTL.

Let $\sum$ be a set of atomic propositional formulas and $p \in \sum$. A CTL formula is defined inductively in the Backus–Naur form as

$$\varphi ::= \top \,|\, \bot \,|\, p \,|\, (\neg\varphi) \,|\, (\varphi \wedge \psi) \,|\, (\varphi \vee \psi) \,|\, (\varphi \to y)$$
$$|(\mathbf{AX}\varphi)|(\mathbf{EX}\varphi)\,|(\mathbf{AG}\varphi)|(\mathbf{EG}\varphi)|(\mathbf{AF}\varphi)|(\mathbf{EF}\varphi)$$
$$|\mathbf{A}(\varphi\mathbf{U}\psi)|\mathbf{E}(\varphi\mathbf{U}\psi) \qquad\qquad\qquad (10.2)$$

In CTL formulas, operators $\neg$, **AG**, **EG**, **AF**, **EF**, **AX**, and **EX** have the highest binding priority, then $\wedge$ and $\vee$, and then $\to$, **AU**, and **EU**. The following formulas are well-formed CTL formulas:

- $\mathbf{AG}(p \wedge q \to \mathbf{EF}\,r)$
- $\mathbf{A}((p \vee q)\,\mathbf{U}\,(\mathbf{EF}\,r))$
- $\mathbf{EF}\,\mathbf{E}(p\,\mathbf{U}\,q)$
- $\mathbf{AG}\,(\mathbf{EF}\,p \to \mathbf{EF}(q \to r\,))$
- $\neg p \wedge r \to \mathbf{EF}(q \to \mathbf{E}(p\,\mathbf{U}\,q))$

```
            P  T
           /  \
there exists a path E    X next state
      for all path A     F future state
                        G globally
                        U until
```

Figure 10.7 CTL operators. P: path quantifiers; T: temporal operators.

The following are not well-formed CTL formulas:

- $AG(p \wedge q \to F r)$
- $(p \vee q) U (EF r)$
- $EF (p U q)$
- $AG (EF p \to G(q \to r))$
- $\neg p \wedge r \to EF(q \to E(p \to q))$

10.2.2.2 Semantics of CTL

CTL formulas are evaluated over state transition systems. Let $M = (S, I, R, L)$ be a transition system, $s \in S$, and φ and ψ be CTL formulas. The satisfaction relation $M, s \vDash \varphi$ is defined as follows:

- $M, s \vDash \top$ and $M, s \nvDash \bot$
- $M, s \vDash p$ iff $p \in L(s)$
- $M, s \vDash \neg\varphi$ iff $M, s \nvDash \varphi$
- $M, s \vDash \varphi \wedge \psi$ iff $M, s \vDash \varphi$ and $M, s \vDash \psi$
- $M, s \vDash \varphi \vee \psi$ iff $M, s \vDash \varphi$ or $M, s \vDash \psi$
- $M, s \vDash \varphi \to \psi$ iff $M, s \vDash \psi$ whenever $M, s \vDash \varphi$
- $M, s \vDash AX \varphi$ iff for all s' such that $(s, s') \in R$, we have $M, s' \vDash \varphi$
- $M, s \vDash EX \varphi$ iff for some s' such that $(s, s') \in R$, we have $M, s' \vDash \varphi$
- $M, s \vDash AG \varphi$ iff for any state s_i along any path we have $M, s_i \vDash \varphi$
- $M, s \vDash EG \varphi$ iff there is a path and for any state s_i along the path we have $M, s_i \vDash \varphi$
- $M, s \vDash AF \varphi$ iff there is a state s_i along every path such that $M, s_i \vDash \varphi$
- $M, s \vDash EF \varphi$ iff there is a path and for some state s_i along the path we have $M, s_i \vDash \varphi$
- $M, s \vDash A(\varphi U \psi)$ iff for all paths $\varphi U \psi$ is satisfied
- $M, s \vDash E(\varphi U \psi)$ there exists a path in which $\varphi U \psi$ is satisfied.

The semantics of **AX** and **EX**, **AG** and **EG**, **AF** and **EF**, and **AU** and **EU** are intuitively illustrated in Figures 10.8–10.11. It also shows that **AF** φ is an abbreviation of $A(\top U \varphi)$ and **EF**φ is an abbreviation of $E(\top U \varphi)$.

Example 10.4 *CTL Formula Verification*
In this example, we verify the system M shown in Example 10.2 against a few CTL specifications from the initial state s_1.

1. $\psi = EX p$.
   $M, s_1 \vDash EX p$ holds because s_2 has p.
2. $\varphi = AF q$.
   Fq is true in every path. Therefore, $M, s_1 \vDash EX q$ holds.
3. $\varphi = EG(\neg p \wedge q)$.
   $G(\neg p \wedge q)$ does not hold in any single path. Therefore, $M, s_1 \vDash EG(\neg p \wedge q)$ does not hold.

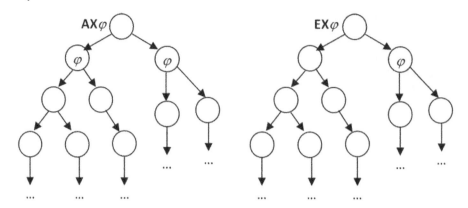

Figure 10.8 Illustration of semantics of AX and EX.

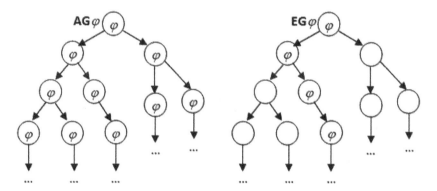

Figure 10.9 Illustration of semantics of AG and EG.

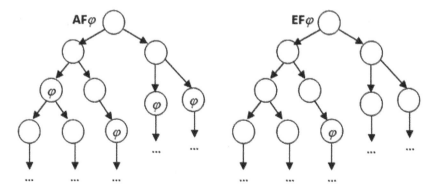

Figure 10.10 Illustration of semantics of AF and EF.

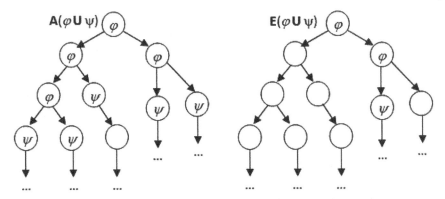

Figure 10.11 Illustration of semantics of AU and EU.

4. $\varphi = \mathbf{A}(q \ \mathbf{U} \ r)$

   We check to see if we have $q \ \mathbf{U} \ r$ along every path. The answer is yes. Therefore, $M, s_1 \vDash \mathbf{A}(q \ \mathbf{U} \ r)$ holds.

5. $\varphi = \mathbf{AF} \ p \to \mathbf{AF} \ q$.

   In general, to verify whether a formula of the form $\mathbf{AF}\psi \to \mathbf{AF}\psi\prime$ is satisfied, we first check to see if $\mathbf{AF}\psi$ is satisfied. If $\mathbf{AF}\psi$ is not satisfied, then $\mathbf{AF}\psi \to \mathbf{AF}\psi'$ is satisfied (recall the truth table of the implication operator). If $\mathbf{AF}\psi$ is satisfied, then we further check to see if $\mathbf{AF}\psi'$ is satisfied. If $\mathbf{AF}\psi'$ is satisfied, then $\mathbf{AF}\psi \to \mathbf{AF}\psi'$ is satisfied. Otherwise, $\mathbf{AF}\psi \to \mathbf{AF}\psi'$ is not satisfied. Because the system shown in Figure 10.5 does not satisfy $\mathbf{AF} \ p$, so $M, s_1 \vDash \mathbf{AF} \ p \to \mathbf{AF} \ q$ holds.

6. $\varphi = \mathbf{AG} \ \mathbf{AF} \ q$.

   In general, to verify whether a formula of the form $\mathbf{AG} \ \mathbf{AF} \ \psi$ is satisfied by a system, first we check to see if $\mathbf{AF} \ \psi$ is satisfied. If $\mathbf{AF} \ \psi$ is not satisfied, then $\mathbf{AG} \ \mathbf{AF}\psi$ is not satisfied. If $\mathbf{AF}\psi$ is satisfied, then we check to see if $\mathbf{AF} \ \psi$ is satisfied from any single state of any single path. If it is, then $\mathbf{AG} \ \mathbf{AF} \ \psi$ is satisfied. In this example, $\mathbf{AF} \ q$ is satisfied because $\mathbf{F} \ q$ is satisfied in every path. In addition, because $\mathbf{AF} \ q$ is satisfied from any state of every path, $M, s_1 \vDash \mathbf{AG} \ \mathbf{AF} \ q$ holds.

7. $\varphi = \mathbf{AX} \ \mathbf{EX} \ q$.

   In general, to verify whether a formula of the form $\mathbf{AX} \ \mathbf{EX}\psi$ is satisfied by a system, we first identify all states s' such that $(s, s') \in R$, and then from each s' we check to see if $\mathbf{EX} \ \psi$ is satisfied. If $\mathbf{EX} \ \psi$ is satisfied from all s', then $\mathbf{AX} \ \mathbf{EX} \ \psi$ holds. In this example, s_2 and s_3 are the two next states of s_1. $\mathbf{EX} \ q$ is satisfied from s_2 but not from s_3. Therefore, $M, s_1 \vDash \mathbf{AX} \ \mathbf{EX} \ q$ does not hold.

8. $\varphi = \mathbf{EX} \ \mathbf{AX} \ q$.

   In general, to verify whether a formula of the form $\mathbf{EX} \ \mathbf{AX}\psi$ is satisfied by a system, we first identify all states s' such that $(s, s') \in R$. Then from each s', we check to see if $\mathbf{AX}\psi$ is satisfied. As long as $\mathbf{AX}\psi$ is satisfied from one s', $\mathbf{EX} \ \mathbf{AX}\psi$ holds. In this example, s_2 and s_3 are the two next states of s_1. $\mathbf{AX} \ q$ is satisfied from none of them. Therefore, $M, s_1 \vDash \mathbf{EX} \ \mathbf{AX} \ q$ does not hold.

10.2.2.3 Equivalencies of CTL Formulas

Two CTL formulas φ and ψ are said to be *semantically equivalent*, denoted by $\varphi \equiv \psi$, if for all models M and all states s in M,

$$M, s \vDash \varphi \text{ iff } M, s \vDash \psi$$

In other words, two CTL formulas are equivalent if either both of them are satisfied or none of them are satisfied from any state of any Kripke structure. A few equivalencies are as follows:

$$\mathbf{AX}\,(\varphi \wedge \psi) \;\equiv\; \mathbf{AX}\;\varphi \wedge \mathbf{AX}\;\psi$$
$$\mathbf{EX}\,(\varphi \vee \psi) \;\equiv\; \mathbf{EX}\;\varphi \vee \mathbf{EX}\;\psi$$
$$\mathbf{AG}\,(\varphi \wedge \psi) \;\equiv\; \mathbf{AG}\;\varphi \wedge \mathbf{AG}\;\psi$$
$$\mathbf{EF}\,(\varphi \vee \psi) \;\equiv\; \mathbf{EF}\;\varphi \vee \mathbf{EF}\;\psi$$
$$\neg\mathbf{AX}\;\varphi \;\equiv\; \mathbf{EX}\;\neg\varphi$$
$$\neg\mathbf{AF}\;\varphi \;\equiv\; \mathbf{EG}\;\neg\varphi$$
$$\neg\mathbf{EF}\;\varphi \;\equiv\; \mathbf{AG}\;\neg\varphi$$
$$\mathbf{AF}\;\mathbf{AF}\;\varphi \;\equiv\; \mathbf{AF}\;\varphi$$
$$\mathbf{EF}\;\mathbf{EF}\;\varphi \;\equiv\; \mathbf{EF}\;\varphi$$
$$\mathbf{AG}\;\mathbf{AG}\;\varphi \;\equiv\; \mathbf{AG}\;\varphi$$
$$\mathbf{EG}\;\mathbf{EG}\;\varphi \;\equiv\; \mathbf{EG}\;\varphi$$

10.2.3 LTL versus CTL

LTL and CTL are two of the most popular forms of temporal logic. LTL views time as a linear path extending to the future, while CTL views time as a branching-out structure. In general, LTL formulas are more intuitive and easier to understand. Using a combination of path quantifiers and temporal operators, CTL formulas are less intuitive and thus more error-prone in system property specification.

They overlap in their expressive powers. CTL allows explicit quantification over paths, which makes it more expressive than LTL in that regard. In fact, any CTL formula necessitating the operator **E** cannot be expressed in LTL.

On the other hand, there are also LTL formulas that cannot be expressed in CTL. One such formula, for example, is the formula $\mathbf{F}\,\varphi \to \mathbf{F}\,\psi$. It means that "all paths that have a state satisfying φ along them also have a state satisfying ψ along them." We might think that it is equivalent to the CTL formula $\mathbf{AF}\,\varphi \to \mathbf{AF}\,\psi$. Actually it is not, because this CTL formula means that "if all paths have a state satisfying φ along them, then all paths have a state satisfying ψ along them." Let $\varphi = p \wedge r$ and $\psi = p \wedge q$, then the model shown in Figure 10.5 does not satisfy the LTL formula $\mathbf{F}\,\varphi \to \mathbf{F}\,\psi$, because φ is satisfied by the left child node of the initial state, but $\mathbf{F}\psi$ is not satisfied along all paths starting from the left child node. However, the model does satisfy the CTL formula **AF** $\varphi \rightarrow$ **AF** ψ because **AF** φ is not satisfied.

There is a temporal logic called CTL* that combines the expressiveness of LTL and CTL. However, it is beyond the scope of this book.

10.3 The NuSMV Model Checking Tool

There is a long list of model checking tools developed to support system property verification. SPIN, NuSMV, FDR2, CADP, and ProB are some examples. They differ in terms of property specification languages and modeling languages. In this book, we only introduce the NuSMV model checker.

NuSMV is a short form for New Symbolic Model Verifier. It is an open-source product jointly developed by ITC-IRST, Trento, Italy, Carnegie Mellon University, the University of Genoa, and the University of Trento. NuSMV supports the analysis of specifications expressed in CTL and LTL. NuSMV is a reimplementation of and extension to SMV, the first model checker based on binary decision diagrams (BDDs).

10.3.1 Description Language

An SMV program is broken down into *modules* that can be composed and reused. Modules describe initial values of variables and how they change in each step. Variables can be of Boolean type, enumerative type, bounded integers, or finite arrays. For example, the following code segment defines a Boolean variable cond, an enumerative variable status that takes values from {ready, busy, waiting, stopped}, an integer variable num that is bounded from 1 to 10, and an array arr of Booleans indexed from 0 to 10.

```
VAR
    cond : Boolean;
    status : {ready, busy, waiting, stopped};
    num : 1..10;
    arr : array 0..10 of Boolean;
```

10.3.1.1 Single-Module SMV Program

Figure 10.12 lists a program of single module. It shows that a module consists of a variable declaration section, started with the keyword VAR, and an assignment section, started with the keyword ASSIGN. In this example, variables are request, a Boolean, and state, an enumerative. The state variables determine the state space of the model.

```
MODULE main
   VAR
      request : boolean;
      state : {ready, busy};
   ASSIGN
      init(state) := ready;
      next(state) :=
         case
            state = ready & request   : busy;
            TRUE                       : {ready, busy};
         esac;
```

Figure 10.12 A single-module SMV program.

The first part of the assignment section assigns the initial value of each variable. The keyword `init` is used to describe the initial value of a variable. The syntax is

```
init(<variable>) := <simple_expression>;
```

where `<simple_expression>` must evaluate to values in the domain of `<variable>`. If the initial value of a variable is not specified, then the variable can take any value in its domain as its initial value. In this example, the variable `state` is initialized to `ready`. The variable `request` is not initialized, and thus, its initial value can be either `TRUE` or `FALSE`. Therefore, there are two initial states in this model:

Initial state 1: `request = TRUE, state = ready;`
Initial state 2: `request = FALSE, state = ready.`

The second part of the assignment section assigns values with the keyword `next`, which describes how the value of the variable changes in one step. The syntax of the `next` statement is

```
next(<variable>) := <next_expression>;
```

where `<next_expression>` must evaluate to values in the domain of `<variable>`. `<next_expression>` depends on "current" and "next" variables. For example,

```
next(x) := x xor TRUE;
next(y) := y & next(x);
```

If the next value is unspecified, then the variable takes any value in its domain at the next step. In this example, the change of `state` is specified with a case statement, but not of `request`. A case statement assigns the value of

Figure 10.13 The transition system corresponding to the SMV program in Figure 10.12.

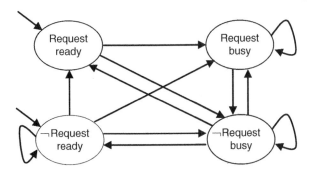

the variable associated with each case condition when it is true; TRUE is for all default cases. In general, it is written as

```
case
    c1 : e1;
    c2 : e2;
    . . .
    TRUE :en;
esac;
```

Let us consider all possible transitions from the first initial state. Based on the first condition of the case statement, the next value of state should be busy. The next value of request is not specified, and thus, it can be either TRUE or FALSE. Therefore, the system can transit from the initial state (request = TRUE, state = ready) to either (request = TRUE, state = busy) or (request = FALSE, state = busy).

Consider the second initial state now. The first condition of the case statement for state is not satisfied, so its next value can be either ready or busy, as defined by the default case. The next value of request again can be either TRUE or FALSE. Therefore, the system can transit from the initial state (request = TRUE, state = ready) to a state of any combination of values of request and state. The state transition model of the program is fully depicted in Figure 10.13.

There are four groups of operators that can be used in SMV expressions:

- Arithmetic operators: +, −, *, /, mod
- Comparison operators: =, !=, >, <, <=, >=
- Logic operators: & (and), | (or), xor (exclusive or), ! (not), → (implication)
- Set operators: in (set inclusion), union (set union)

10.3.1.2 Multimodule SMV Program
An SMV program can consist of more than one module. In each SMV specification, there must be a module main. It is the top-most module. All other

```
MODULE counter_cell(carry_in)
    VAR
        value : boolean;
    ASSIGN
        init(value) := FALSE;
        next(value) := value xor carry_in;
    DEFINE carry_out := value & carry_in;

MODULE main
    VAR
        bit0 : counter_cell(TRUE);
        bit1 : counter_cell(bit0.carry_out);
        bit2 : counter_cell(bit1.carry_out);
```

Figure 10.14 SMV program of a three-bit counter.

modules are instantiated in main or other parent modules. The instantiation is performed inside the VAR declaration of the parent module. All the variables declared in a module instance are visible in the module in which it has been instantiated via the dot notation.

The program listed in Figure 10.14 has two modules: main and counter_cell. It is a model of a three-bit binary counter circuit. As the name indicates, the entry module of the program is main. The main module simply initiates three instances of the counter_cell module, named bit0, bit1, and bit2. The counter_cell module has a parameter carry_in. For example, the carry_in of bit1 is bit0.carry_out. Note that an expression of the form a.b denotes the component b of module a, just as if the module a were a data structure in a standard programming language. Hence, the carry_in of module bit1 is the carry_out of module bit0. DEFINE is used to define C-like "macros"; defined variables are not real variables in that they do not increase the state space. The operator xor means "exclusive or," a Boolean operator working on two variables that has the value of 1 (TRUE) if one but not both of the variables has a value of 1 (TRUE).

The initial state of the model is

```
(bit0.value = FALSE, bit0.value = FALSE,
    bit0.value = FALSE)
```

or briefly described as 000. It transitions to 001, 010, ... and all the way to 111 and then repeats. The details of the first two transitions are listed in Table 10.2. The result of each transition is recorded in next(value).

Table 10.2 First two state transitions of the program listed in Figure 10.14.

	Transition 1			Transition 2		
	bit0	bit1	bit2	bit0	bit1	bit2
carry_in	T	F	F	T	T	F
value	F	F	F	T	F	F
next(value)	T	F	F	F	T	F
carry_out	F	F	F	T	F	F

10.3.1.3 Asynchronous Systems

The previous two programs describe synchronous systems, where in each module or module instance, the assignment statements are taken into account in parallel and simultaneously at each "clock tick." NuSMV allows for asynchronous system modeling. It is possible to define a collection of parallel processes, whose actions are interleaved, following an asynchronous model of concurrency.

Figure 10.15 lists an SMV program that represents a ring of three asynchronous inverting gates. Here, the key word `process` specifies asynchronous module instances. Each time the global clock ticks, only one of the three `inverter` instances is randomly chosen to execute, and the values of variables of other instances remain unchanged. Since the system is not forced to eventually choose a given process to execute, it is possible that the output

```
MODULE inverter(input)
    VAR
        output : boolean;
    ASSIGN
        init(output) := FALSE;
        next(output) := !input;

MODULE main
VAR
        gate1 : process inverter(gate3.output);
        gate2 : process inverter(gate1.output);
        gate3 : process inverter(gate2.output);
    FAIRNESS
        running
```

Figure 10.15 SMV program of an inverter ring.

of a given gate may remain constant forever, regardless of its input. Thus, a statement

```
FAIRNESS
    running
```

is added to the end of the inverter module to force every instance of inverter to execute infinitely often.

10.3.2 Specifications

Specifications can be added in any module of an SMV program. Each property is verified separately. NuSMV supports specifications in LTL and CTL. A property in LTL is specified with the keyword LTLSPEC:

```
LTLSPEC <ltl_expr>
```

where <ltl_expr> is an LTL formula coded in NuSMV:

```
ltl_expr ::
    simple_expr ;; a simple boolean expression
  | "(" ltl_expr ")"
  | "!" ltl_expr ;; logical not
  | ltl_expr "&" ltl_expr ;; logical and
  | ltl_expr "|" ltl_expr ;; logical or
  | ltl_expr "xor" ltl_expr ;; logical exclusive or
  | ltl_expr "->" ltl_expr ;; logical implies
  | ltl_expr "<->" ltl_expr ;; logical equivalence ;;
  | "X" ltl_expr ;; next state
  | "G" ltl_expr ;; globally
  | "F" ltl_expr ;; finally
  | ltl_expr "U" ltl_expr ;; until
  | ltl_expr "V" ltl_expr ;; releases
```

For example, we can add

```
LTLSPEC F (bit0.value & bit1.value & bit2.value)
```

to the end of the program listed in Figure 10.14. This specification checks whether the property that eventually the counter outputs 111 holds. For the same program, we can also add

```
LTLSPEC G F bit2.value
```

which checks whether the third bit becomes true infinitely often.

A property in CTL is specified with the keyword SPEC:

```
SPEC <ctl_expr>
```

where <ctl_expr> is a CTL formula coded in NuSMV:

```
ctl_expr ::
    simple_expr ;; a simple boolean expression
    | "(" ctl_expr ")"
    | "!" ctl_expr ;; logical not
    | ctl_expr "&" ctl_expr ;; logical and
    | ctl_expr "|" ctl_expr ;; logical or
    | ctl_expr "xor" ctl_expr ;; logical exclusive or
    | ctl_expr "->" ctl_expr ;; logical implies
    | ctl_expr "<->" ctl_expr ;; logical equivalence
    | "EG" ctl_expr ;; exists globally
    | "EX" ctl_expr ;; exists next state
    | "EF" ctl_expr ;; exists finally
    | "AG" ctl_expr ;; forall globally
    | "AX" ctl_expr ;; forall next state
    | "AF" ctl_expr ;; forall finally
    | "E" "[" ctl_expr "U" ctl_expr "]" ;; exists until
    | "A" "[" ctl_expr "U" ctl_expr "]" ;; forall until
```

For example, we can add

```
SPEC EX gate3.output
SPEC EX gate1.output -> EX gate2.output
SPEC EF ((!gate1.output) & (!gate2.output) &
    gate3.output)
```

to the end of the program listed in Figure 10.15.

In addition to the properties specified in LTL or CTL formulas, NuSMV can verify the *invariant* properties. An invariant is a propositional property, which must always hold and specified using the keyword INVARSPEC:

```
INVARSPEC <simple_expression>
```

For example, for the program listed in Figure 10.12, we add the following specification:

```
INVARSPEC state in {ready, busy}
```

to check to see if the variable state always has a legal value.

10.3.3 Running NuSMV

NuSMV can be used either interactively or in batch mode. To check a model against a set of specifications, we write the specification and system description in a file with .smv extension and type the command

```
NuSMV <file_name>.smv
```

NuSMV will check each specification automatically, informing whether it is satisfied, or will produce a trace (when possible) to demonstrate its violation. For example, we save the counter program with the two LTL specifications we mentioned earlier to a file named counter.smv and run the command

```
NuSMV counter.smv
```

We will get the following result:

```
*** This version of NuSMV is linked to the MiniSat
    SAT solver.
*** See http://minisat.se/MiniSat.html
*** Copyright (c) 2003-2006, Niklas Een,
    Niklas Sorensson
*** Copyright (c) 2007-2010, Niklas Sorensson
-- specification  F ((bit0.carry_out & bit1.carry_out)
   & bit2.carry_out)  is true
-- specification  G ( F bit2.value)  is true
```

If we modify the second specification to

```
LTLSPEC G X bit2.value
```

and run the command again, we get

```
*** This version of NuSMV is linked to the MiniSat
    SAT solver.
*** See http://minisat.se/MiniSat.html
*** Copyright (c) 2003-2006, Niklas Een,
    Niklas Sorensson
*** Copyright (c) 2007-2010, Niklas Sorensson
-- specification  F ((bit0.value&bit1.value)
   &bit2.value)  is true
-- specification  G ( X bit2.carry_out)  is false
-- as demonstrated by the following execution sequence
Trace Description: LTL Counterexample
Trace Type: Counterexample
  -- Loop starts here
  -> State: 1.1 <-
    bit0.value = FALSE
    bit1.value = FALSE
    bit2.value = FALSE
    bit0.carry_out = FALSE
    bit1.carry_out = FALSE
```

```
        bit2.carry_out = FALSE
    -> State: 1.2 <-
        bit0.value = TRUE
        bit0.carry_out = TRUE
    -> State: 1.3 <-
        bit0.value = FALSE
        bit1.value = TRUE
        bit0.carry_out = FALSE
    -> State: 1.4 <-
        bit0.value = TRUE
        bit0.carry_out = TRUE
        bit1.carry_out = TRUE
    -> State: 1.5 <-
        bit0.value = FALSE
        bit1.value = FALSE
        bit2.value = TRUE
        bit0.carry_out = FALSE
        bit1.carry_out = FALSE
    -> State: 1.6 <-
        bit0.value = TRUE
        bit0.carry_out = TRUE
    -> State: 1.7 <-
        bit0.value = FALSE
        bit1.value = TRUE
        bit0.carry_out = FALSE
    -> State: 1.8 <-
        bit0.value = TRUE
        bit0.carry_out = TRUE
        bit1.carry_out = TRUE
        bit2.carry_out = TRUE
    -> State: 1.9 <-
        bit0.value = FALSE
        bit1.value = FALSE
        bit2.value = FALSE
        bit0.carry_out = FALSE
        bit1.carry_out = FALSE
        bit2.carry_out = FALSE
NuSMV >
```

The result indicates that the second specification is false, and it prints out a counter example of execution that violates the specification. The counter example lists all states step by step until they repeat.

NuSMV supports simulation that allows users to explore the possible executions (*traces* from now on) of an SMV model. A simulation session is started interactively from the system prompt as follows:

```
system_prompt> NuSMV -int <file>.smv
NuSMV> go
NuSMV>
```

The next step is to pick a state from the initial states to start a new trace. To pick a state randomly, type command

```
NuSMV> pick_state -r
```

Subsequent states in the simulation can be picked using the `simulate` command. For example, we can type the command

```
NuSMV> simulate -r -k 5
```

to randomly simulate five steps of a trace. To show the trace with states, use the commands

```
NuSMV> show_trace -t
NuSMV> show_trace -v
```

Following is a screenshot of NuSMV simulation:

```
C:\...\NuSMV-2.6.0-win64\bin>nusmv -int counter.smv
...
NuSMV > go
NuSMV > pick_state -r
NuSMV > simulate -r -k 3
********  Simulation Starting From State 2.1  ********
NuSMV > show_traces -t
There are 2 traces currently available.
NuSMV > show_traces -v
    <!-- ################# Trace number: 2
         ################# -->
Trace Description: Simulation Trace
Trace Type: Simulation
  -> State: 2.1 <-
    bit0.value = FALSE
    bit1.value = FALSE
    bit2.value = FALSE
    bit0.carry_out = FALSE
    bit1.carry_out = FALSE
    bit2.carry_out = FALSE
```

```
-> State: 2.2 <-
  bit0.value = TRUE
  bit1.value = FALSE
  bit2.value = FALSE
  bit0.carry_out = TRUE
  bit1.carry_out = FALSE
  bit2.carry_out = FALSE
-> State: 2.3 <-
  bit0.value = FALSE
  bit1.value = TRUE
  bit2.value = FALSE
  bit0.carry_out = FALSE
  bit1.carry_out = FALSE
  bit2.carry_out = FALSE
-> State: 2.4 <-
  bit0.value = TRUE
  bit1.value = TRUE
  bit2.value = FALSE
  bit0.carry_out = TRUE
  bit1.carry_out = TRUE
  bit2.carry_out = FALSE
  bit2.carry_out = FALSE
NuSMV >
```

For details of how to use the NuSMV tool, please read the latest version of NuSMV tutorial that can be downloaded from the NuSMV official website.

10.4 Real-Time Computation Tree Logic

RTCTL is a real-time extension to CTL, where operators **G**, **F**, and **U** are bounded. RTCTL specifies temporal properties not only qualitatively but also quantitatively.

Let $\sum$ be a set of atomic propositional formulas and $p \in \sum$. In addition, let k be a natural number. An RTCTL formula is defined inductively in the Backus–Naur form as

$$\varphi ::= \mathsf{T} \mid \perp \mid p \mid (\neg\varphi) \mid (\varphi \wedge \psi) \mid (\varphi \vee \psi) \mid (\varphi \rightarrow \psi)$$

$$\mid (\mathbf{AX}\varphi) \mid (\mathbf{EX}\varphi) \mid (\mathbf{AG}\varphi) \mid (\mathbf{EG}\varphi) \mid (\mathbf{AF}\varphi) \mid (\mathbf{EF}\varphi) \mid \mathbf{A}(\varphi\mathbf{U}\psi) \mid \mathbf{E}(\varphi\mathbf{U}\psi)$$

$$\mid (\mathbf{AG}^{\leq k}\varphi) \mid (\mathbf{EG}^{\leq k}\varphi) \mid (\mathbf{AF}^{\leq k}\varphi) \mid (\mathbf{EF}^{\leq k}\varphi) \mid \mathbf{A}(\varphi\mathbf{U}^{\leq k}\psi) \mid \mathbf{E}(\varphi\mathbf{U}^{\leq k}\psi)$$

$$(10.3)$$

Let $M = (S, I, R, L)$ be a transition system, $s \in S$, and φ and ψ be RTCTL formulas. A path π is denoted as

$$\pi = s_1 \rightarrow s_2 \rightarrow \cdots$$

The satisfaction relation $M, s \vDash \varphi$ for all operators that appear in the CTL definition is exactly the same as that defined in CTL. For those newly added operators, the relationship is defined as follows:

- $M, s_1 \vDash \mathbf{AG}^{\leq k}\varphi$ iff for any state s_i, $i \leq k+1$, along any path π we have $M, s_i \vDash \varphi$.
- $M, s_1 \vDash \mathbf{EG}^{\leq k}\varphi$ iff there is a path π such that for any state s_i, $i \leq k+1$, along the path we have $M, s_i \vDash \varphi$
- $M, s_1 \vDash \mathbf{AF}^{\leq k}\varphi$ iff along any path π, there is a state s_i, $i \leq k+1$, such that $M, s_i \vDash \varphi$.
- $M, s_1 \vDash \mathbf{EF}^{\leq k}\varphi$ iff there is a path π and there is a state s_i, $i \leq k+1$, along this path such that $M, s_i \vDash \varphi$.
- $M, s_1 \vDash \mathbf{A}(\varphi\, \mathbf{U}^{\leq k}\psi)$ iff for any path π, there exists a state s_i, $0 \leq i \leq k$, such that $M, s_i \vDash \psi$ and for any j, $0 \leq j < i$, $M, s_j \vDash \varphi$.
- $M, s_1 \vDash \mathbf{E}(\varphi\, \mathbf{U}^{\leq k}\psi)$ iff for some path π, there exists an i, $0 \leq i \leq k$, such that $M, s_i \vDash \psi$ and for any j, $0 \leq j < i$, $M, s_j \vDash \varphi$.

RTCTL is useful in specifying real-time system properties. For example, we can specify the maximal temporal distance between two events A and B as follows:

$\mathbf{AG}\,(\mathrm{A} \rightarrow \mathbf{AF}^{\leq k}\; \mathrm{B})$.

If $k = 3$, this RTCTL formula is equal to the following CTL formula:

$\mathbf{AG}\,(\mathrm{A} \rightarrow (\mathrm{B} \vee \mathbf{AX}\;(\mathrm{B} \vee \mathbf{AX}\;(\mathrm{B} \vee \mathbf{AX}\; \mathrm{B}))))$

Of course, if k is large, such a translation will result in an exponential blowup in CTL formulas. The exact temporal distance between two events can be specified as

$\mathbf{AG}\;(\mathrm{A} \rightarrow (\mathbf{AG}^{\leq\, k-1}\; \neg \mathrm{B} \wedge \mathbf{AF}^{\leq\, k}\; \mathrm{B}))$

The minimal temporal distance between two consecutive occurrences of an event can be specified as

$\mathbf{AG}\;(\mathrm{E} \rightarrow \mathbf{AG}^{\leq k}\; \neg \mathrm{E})$.

To specify the periodicity of a task, we can use the following formula:

$\mathbf{AG}\;(\mathrm{E} \rightarrow (\mathbf{AG}^{\leq k-1}\; \neg \mathrm{E} \wedge \mathbf{AF}^{\leq k}\; \mathrm{E}))$

NuSMV allows RTCTL specifications. RTCTL extends the syntax of CTL path expressions with the following bounded modalities:

```
rtctl_expr :: ctl_expr
       | EBF range rtctl_expr
       | ABF range rtctl_expr
       | EBG range rtctl_expr
       | ABG range rtctl_expr
       | A [ rtctl_expr BU range rtctl_expr ]
       | E [ rtctl_expr BU range rtctl_expr ]
range :: integer_number .. integer_number
```

For details of RTCTL expressions and RTCTL specifications in NuSMV, please read the latest version of the NuSMV User Manual.

Example 10.5 *The Ferryman Puzzle*

In the Exercise section of Chapter 9, we mentioned the ferryman puzzle. The puzzle can be solved with NuSMV model checker.

To model the system with a NuSMV program, we ignore the boat, as it is always with the ferryman. The four agents, the ferryman, wolf, goat, and cabbage, are modeled with four Boolean variables, as they each have two values: *false* (on the initial bank of the river) and *true* (on the destination bank of the river). We model all possible behaviors in the program and ask if a trace indicating that the ferryman takes all of the three passengers to the other bank safely exists. The program is listed in Figure 10.16. In the program, we use the variable `carry` to indicate which passenger the ferryman will take with him to cross the river. If the ferryman travels alone, then `carry` takes the value of 0.

In the ASSIGN section of the program, the statement

```
next(ferryman) := !ferryman;
```

means that the ferryman must cross the river in each step. The `next(carry)` statement shows that the ferryman can take any passenger that is on the same bank of the river as he is or takes nothing with him (`union` 0 at the end of the statement). The `next(goat)` statement says that if the goat and ferryman are on the same bank and the goat is chosen to cross the river with the ferryman, then the goat will be on the other bank in the next state. Otherwise, the goat will stay on the same bank. The `next(cabbage)` and `next(wolf)` statements are similar to the `next(goat)` statement.

The LTL specification states that for all execution paths, if the goat and cabbage are on the same bank, or if the goat and wolf are on the same bank, then

```
MODULE main
   VAR
      ferryman : boolean;
      goat     : boolean;
      cabbage  : boolean;
      wolf     : boolean;
      carry    : {g, c, w, 0};
   ASSIGN
      init(ferryman) := FALSE;
      init(goat)     := FALSE;
      init(cabbage)  := FALSE;
      init(wolf)     := FALSE;
      init(carry)    := 0;

      next(ferryman) := !ferryman;

      next(carry) :=
         case
            (ferryman = goat) : g;
            TRUE              : 0;
         esac union
         case
            (ferryman = cabbage): c;
            TRUE                : 0;
         esac union
         case
            (ferryman = wolf)  : w;
            TRUE               : 0;
         esac union 0;

      next(goat) :=
         case
            (ferryman = goat) & (next(carry) = g)
                              : next(ferryman);
               TRUE           : goat;
         esac;

      next(cabbage) :=
         case
            (ferryman = cabbage) & (next(carry) = c)
                              : next(ferryman);
               TRUE           : cabbage;
         esac;
```

Figure 10.16 SMV program of the ferryman puzzle.

```
next(wolf) :=
    case
        (ferryman = wolf) & (next(carry) = w)
                                : next(ferryman);
            TRUE                : wolf;
    esac;
```

```
LTLSPEC
((goat=cabbage |goat = wolf) -> goat = ferryman)
        U (cabbage & goat & wolf & ferryman)
```

Figure 10.16 (*Continued*)

the goat must be with the ferryman, and this is true until the ferryman and all of his passengers are on the destination bank. The program produces the following output:

```
C:\...\NuSMV-2.6.0-win64\bin>nusmv ferryman.smv
...
-- specification (((goat = cabbage | goat = wolf) ->
    goat = ferryman) U (((cabbage & goat) & wolf) &
    ferryman))  is false
-- as demonstrated by the following execution sequence
Trace Description: LTL Counterexample
Trace Type: Counterexample
  -- Loop starts here
  -> State: 1.1 <-
    ferryman = FALSE
    goat = FALSE
    cabbage = FALSE
    wolf = FALSE
    carry = 0
  -> State: 1.2 <-
    ferryman = TRUE
  -> State: 1.3 <-
    ferryman = FALSE
```

The result shows that the property is false, which is correct, because there are plenty of "execution paths" that violate this property. The program prints out one example, in which the ferryman crosses the river alone first (state 1.2), then he travels back (state 1.3), and he keeps crossing the river this way.

Because what we are looking for is an execution path such that the LTL specification is true, we can verify an opposite specification, which is

```
LTLSPEC
!(((goat=cabbage |goat = wolf) -> goat = ferryman)
     U (cabbage & goat & wolf & ferryman))
```

If the puzzle has a solution, then the aforementioned specification won't be true on all execution paths. In that case, the trace of a counter example will be printed on running the program, which is the solution that we want. The result from running the program with the new specification is listed as follows, in which states 1.1 through 1.8 are the trace of the solution, while states 1.9 through 1.15 are the trace that the ferryman safely takes all passengers back from the destination bank to the original bank.

```
C:\...\NuSMV-2.6.0-win64\bin>nusmv ferryman.smv
...
-- specification !(((goat = cabbage | goat = wolf) ->
   goat = ferryman) U (((cabbage & goat) & wolf) &
   ferryman)) is false
-- as demonstrated by the following execution sequence
Trace Description: LTL Counterexample
Trace Type: Counterexample
  -- Loop starts here
  -> State: 1.1 <-
    ferryman = FALSE
    goat = FALSE
    cabbage = FALSE
    wolf = FALSE
    carry = 0
  -> State: 1.2 <-
    ferryman = TRUE
    goat = TRUE
    carry = g
  -> State: 1.3 <-
    ferryman = FALSE
    carry = 0
  -> State: 1.4 <-
    ferryman = TRUE
    wolf = TRUE
    carry = w
  -> State: 1.5 <-
    ferryman = FALSE
    goat = FALSE
```

```
   carry = g
-> State: 1.6 <-
   ferryman = TRUE
   cabbage = TRUE
   carry = c
-> State: 1.7 <-
   ferryman = FALSE
   carry = 0
-> State: 1.8 <-
   ferryman = TRUE
   goat = TRUE
   carry = g
-> State: 1.9 <-
   ferryman = FALSE
   wolf = FALSE
   carry = w
-> State: 1.10 <-
   ferryman = TRUE
   carry = 0
-> State: 1.11 <-
   ferryman = FALSE
   cabbage = FALSE
   carry = c
-> State: 1.12 <-
   ferryman = TRUE
   carry = 0
-> State: 1.13 <-
   ferryman = FALSE
   goat = FALSE
   carry = g
-> State: 1.14 <-
   ferryman = TRUE
   carry = 0
-> State: 1.15 <-
   ferryman = FALSE
```

Exercises

1　What is model checking? Why do we need model checking?

2　What is the difference between LTL and CLT in terms of modeling of time?

3   Can any LTL formula be specified by CTL? Can any CTL formula be specified by LTL?

4   What is the difference between CTL and RTCTL?

5   Draw the parse trees for the following LTL formulas:
(a) $\neg p \wedge \mathbf{X} p \rightarrow \mathbf{F} q$
(b) $(p \vee \neg q) \vee \mathbf{X} r \rightarrow \mathbf{GF} q$
(c) $(p \mathbf{U} q) \vee (\mathbf{F} q \wedge \mathbf{G} r)$
(d) $\mathbf{X}(p \vee q) \mathbf{U} (\mathbf{F} q \wedge \neg r)$

6   Draw parse trees for the following CTL formulas:
(a) $\mathbf{EG}((\neg p \wedge \mathbf{EX} p) \rightarrow \mathbf{AG} q)$
(b) $\mathbf{A}(\neg p \mathbf{U} q) \vee (\mathbf{AF} q \wedge \mathbf{EG}(p \vee r))$
(c) $\mathbf{E}(\neg p \mathbf{U} (\mathbf{AG} q)) \rightarrow \mathbf{AG} (q \rightarrow r)$
(d) $\mathbf{AG} \, \mathbf{AF}(\neg p \mathbf{U} (q \vee r))$

7   Consider the state transition system M illustrated in Figure 10.17. For each LTL formula φ listed, decide if $M, s_1 \vDash \varphi$ holds.
(a) $p \wedge q$
(b) $(p \vee q) \wedge \mathbf{X} q$
(c) $\mathbf{X} \, \mathbf{X} q$
(d) $\mathbf{X} \, \mathbf{X}(q \vee r)$
(e) $\mathbf{F} \, q$
(f) $\mathbf{F} \, (p \wedge q)$
(g) $\mathbf{F} \, (q \wedge r)$
(h) $\mathbf{G} \, (p \vee q)$
(i) $\mathbf{G} \, (p \wedge q)$
(j) $p \, \mathbf{U} \, q$
(k) $p \, \mathbf{U} \, r$
(l) $p \, \mathbf{U} \, (q \wedge r)$
(m) $\mathbf{G} \, (p \, \mathbf{U} \, (q \wedge r))$
(n) $p \rightarrow \neg q$
(o) $\mathbf{G} \, (p \rightarrow \neg q)$
(p) $\mathbf{GF} \, r$
(q) $\mathbf{G}(p \rightarrow \mathbf{X} \, r)$
(r) $\mathbf{F}(q \rightarrow \neg r)$
(s) $\mathbf{GF} \, (q \rightarrow r)$
(t) $\mathbf{F} \, (p \wedge q) \rightarrow \mathbf{F} \, (q \wedge r)$
(u) $\mathbf{GF} \, (p \wedge q) \rightarrow \mathbf{GF} \, \neg(q \vee r)$
(v) $\mathbf{G}(p \rightarrow \mathbf{X} \, r)$
(w) $\mathbf{F}(q \rightarrow \neg r)$
(x) $\mathbf{GF}(q \rightarrow r)$

Figure 10.17 The state transition model for Problems 7 and 8.

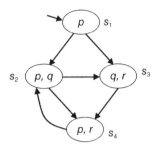

8 Consider the state transition system M illustrated in Figure 10.17. For each CTL formula φ listed, decide if M, $s_1 \vDash \varphi$ holds.
  (a) **AF**$(p \wedge q)$
  (b) **AG**$(p \vee r)$
  (c) **AX AX**q
  (d) **AX EX**q
  (e) **AF**$(q \wedge r)$
  (f) **EF**$(q \wedge r)$
  (g) **AG**$(p \wedge \neg(q \vee r))$
  (h) **EG**$(p \wedge \neg(q \vee r))$
  (i) **E**$(p \textbf{ U } (q \wedge r))$
  (j) **A**$(p \textbf{ U } (q \wedge r))$
  (k) **A**$(p \textbf{ U AG } q)$
  (l) **E**$(p \textbf{ U EG } q)$
  (m) **EG**$(p \rightarrow \neg q)$
  (n) **EG AF** r
  (o) **AG EF** r
  (p) **EG**$(p \rightarrow \textbf{X } r)$
  (q) **EF**$(q \rightarrow \neg r)$
  (r) **GF**$(q \rightarrow r)$

9 Consider the state transition system M illustrated in Figure 10.18.
  (1) Write the SMV program of the model.
  (2) For each of the LTL or CTL formula φ listed, verify if M, $s_1 \vDash \varphi$ holds by running the program.
    (a) $(p \vee q) \wedge \textbf{X } q$
    (b) **F** $(p \vee q)$
    (c) **F G** q
    (d) **G F** q
    (e) **X** $q \rightarrow (p \vee q) \textbf{ U } r$
    (f) **AX** r
    (g) **EX** q

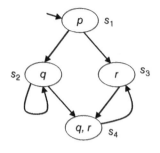

Figure 10.18 The state transition model for Problem 9.

    (h) **EG** r
    (i) **EF AG** r
    (j) **A** $(p\ \mathbf{U}\ r)$
    (k) **EX** $q \to \mathbf{E}((p \vee q)\ \mathbf{U}\ r)$

10 The SMV code listed in Figure 10.19 models the mutual exclusion feature of two processes. Each process can be in one of the three states: *idle, ready* (ready to enter the critical section), or *critical* (in the critical section). The properties under concern are as follows:

Safety. At any moment, at most one task can be in its critical section.

Liveness. Whenever a task requests to enter its critical section (and thus access the shared resource), the request will eventually be granted.

Fairness. If a task makes infinitely often requests to enter its critical section, it will enter its critical section infinitely often.

    (1) Run the program and verify the properties of safety and liveness.
    (2) Code the fairness property in LTL and verify it.
    (3) Draw the state transition diagram of the program.
    (4) Modify the program to include a third process `pr2` such that all the three properties are satisfied.

11 Figure 10.20 shows the control of a microwave oven in terms of its working status.
    (1) Write the SMV program.
    (2) Some fundamental requirements over the oven control include the following:
        (a) The oven cannot heat unless the door is closed.
        (b) No matter what status of the oven is in, it will be heating eventually.
    Code the two properties in CTL and verify them with the SMV program.

```
MODULE main
   VAR
      turn: {0, 1};
      pr0: process prc(pr1.control, turn, 0);
      pr1: process prc(pr0.control, turn, 1);

   ASSIGN
      init(turn) := 0;

      -- safety
   SPEC AG !((pr0.control = critical)&(pr1.control = critical))
      -- liveness
   SPEC AG ((pr0.control = ready) -> AF(pr0.control = critical))
   SPEC AG ((pr1.control = ready) -> AF(pr1.control = critical))


MODULE prc(other_control, turn, ID)
   VAR
      control: {idle, ready, critical};

   ASSIGN
      init(control) := idle;

      next(control) :=
         case
            (control = idle) : {ready, idle};
            (control = ready)&(other_control = idle): critical;
            (control = ready)&(other_control = ready)
                    &(turn = ID)        : critical;
            (control = critical)   : {critical, idle};
            TRUE                        : control;
         esac;

      next(turn) :=
         case
            (turn = ID)&(control = critical) : (turn + 1) mod 2;
            TRUE : turn;
         esac;

FAIRNESS running;
FAIRNESS !(control = critical);
```

Figure 10.19 SMV program of mutual exclusion.

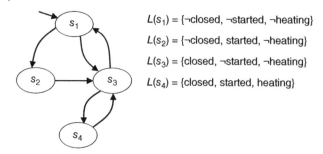

$L(s_1) = \{\neg\text{closed}, \neg\text{started}, \neg\text{heating}\}$

$L(s_2) = \{\neg\text{closed}, \text{started}, \neg\text{heating}\}$

$L(s_3) = \{\text{closed}, \neg\text{started}, \neg\text{heating}\}$

$L(s_4) = \{\text{closed}, \text{started}, \text{heating}\}$

Figure 10.20 State transition system of a microwave oven.

Suggestions for Reading

Burstall [1], Kroger [2], and Pnueli [3], all proposed using temporal logic for reasoning about computer programs. Lamport was the first to investigate the expressive power of various temporal logics for verification. He discussed two logics: a simple linear-time logic and a simple branching-time logic in Ref. [4] and showed that each logic could express certain properties that could not be expressed in the other. The model checking approach to program verification was first proposed by Clarke and Emerson [5–7]. RTCTL was introduced in Ref. [8] by Emerson, Mok, Sistla, and Srinivasan.

Bobbio and Horvath presented a technique to check if a time Petri net (TPN) satisfies the temporal properties expressed in RTCTL [9].The transition graph of the TPN is built in a compositional manner based on discretization of the firing intervals. The compositional description can be automatically translated into the model description language of NuSMV, a tool for model checking finite-state systems against specifications in RTCTL.

Another time extension to temporal logic, called Timed Computation Tree Logic (TCTL), was introduced in Ref. [10]. Virbitskaite and Pokozy [11] proposed a method to model check TCTL properties on TPN.

References

1 Burstall, R.M. (1974) Program proving as hand simulation with a little induction. *IFIP Congress 74*, North Holland, pp. 308–312.
2 Kroger, F. (1977) Lar: A logic of algorithmic reasoning. *Acta Informatica*, **8**, 243–266.
3 Pnueli, A. (1977) The temporal semantics of concurrent programs. *18th Annual Symposium on Foundations of Computer Science*.
4 Lamport, L. (1980) "Sometimes" is sometimes "Not Never". Annual ACM Symposium on Principles of Programming Languages, pp. 174–185.

5 Clarke, E.M. and Emerson, E.A. (1981) Design and synthesis of synchronization skeletons using branching-time temporal logic. *Logic of Programs*, **131**, 52–71.

6 Clarke, E.M., Emerson, E.A., and Sistla, A.P. (1986) Automatic verification of finite-state concurrent systems using temporal logic specifications. *ACM Transactions on Programming Languages and Systems*, **8** (2), 244.

7 Emerson, E.A. and Clarke, E.A. (1980) Characterizing correctness properties of parallel programs using fixpoints. *Proceedings of the 7th Colloquium on Automata, Languages and Programming*, Noordwijkerhout, The Netherlands, 169–181.

8 Emerson, E.A., Mok, A.K., Sistla, A.P., and Srinivasan, J. (1992) Quantitative temporal reasoning. *Real-Time Systems*, **4** (4), 331–352.

9 Bobbio, A. and Horvath, A. (2001) Modeling checking time Petri nets using NuSMV. *Proceedings of the 5th International Workshop on Performability Modeling of Computer and Communication Systems*, pp. 100–104.

10 Alur, R., Courcoubetis, C. and Dill, D. (1990) Model-checking for real-time systems. *Proceedings of the 5th Annual Symposium on Logic in Computer Science*, pp. 414–425.

11 Virbitskaite, I. and Pokozy, E. (1999) A partial order method for the verification of time Petri nets, in *Fundamentals of Computation Theory* (eds G. Ciobanu and G. Paum), LNCS 1684, Springer-Verlag.

11

Practical Issues

This chapter briefly introduces some practical issues in real-time embedded system design and development that designers should be aware of. These issues include software reliability, software aging and rejuvenation, software security, embedded system safety, and power efficiency.

11.1 Software Reliability

Reliability is a measurement of the probability that a system operates without failure over a specified time within a specified environment for a specified purpose. Software can fail. A failure corresponds to unexpected runtime behavior observed by a user of the software. Software failures may be due to errors, ambiguities, oversights or misinterpretation of the specification that the software is supposed to satisfy, carelessness or incompetence in writing code, inadequate testing, incorrect or unexpected usage of the software, or other unforeseen problems. A failure is caused by a fault (or bug) in the code.

11.1.1 Software Faults

A software fault is either a *Bohrbug* or a *Mandelbug*.

Bohrbugs. Bohrbugs are *solid* software faults that can be easily detected and fixed, and the corresponding failure occurrences can be easily reproduced. Bohrbugs are essentially permanent design faults and hence almost deterministic in nature. They can be identified and removed during the testing and debugging phase (or early deployment phase) of the software life cycle.

Mandelbugs. Mandelbugs are bugs whose underlying causes are so complex and obscure that their behaviors appear chaotic and even nondeterministic. They are essentially permanent faults whose conditions of activation occur rarely or are not easily reproducible. The complexity can take two forms.

Real-Time Embedded Systems, First Edition. Jiacun Wang.
© 2017 John Wiley & Sons, Inc. Published 2017 by John Wiley & Sons, Inc.

(i) The activation and/or error propagation depend on a combination of conditions within both the software and its running environment, such as interactions of the system with its environment, the timing of inputs, and operation sequencing. (ii) There is a delay from fault activation to failure occurrence. In general, Mandelbugs are difficult to locate, because the faults and failures might not be near the actual fault activation in code/operation location or time.

There are two special types of Mandelbugs: Heisenbugs and aging-related bugs.

Heisenbugs. Heisenbugs are faults that will stop causing failures or manifest differently when one attempts to probe or isolate them. For example, some failures are related to improper initialization. When debuggers that initialize unused memory are turned on, these failures may disappear.

Aging-related bugs. An aging-related bug is a fault that leads to the accumulation of errors either within the running application or in its system-internal environment, resulting in an increased failure rate and/or degraded performance. Software aging will be discussed in detail in the next section.

Most studies on failure data have reported that a large proportion of software failures are transient in nature. Not every fault causes a failure. There is a famous 90–10 rule, which says that 90% of the time a software application is typically executing 10% of the code. Because of that, many faults simply reside in the software silently and do not cause trouble over a long period of time. This also means that fixing certain percent of faults in a software application does not necessarily improve the application's reliability by the same percentage. One study shows that removing 60% of software defaults only led to a 3% of reliability improvement.

11.1.2 Reliability Measurement

Software reliability is usually measured in terms of *Mean Time between Failures* (MTBF). For example, if MTBF = 5000 hours for average software, then the software is expected to work for 5000 hours for continuous operations.

For safety-critical real-time embedded systems, the reliability can also be measured using *Probability of Failure on Demand* (POFOD), which is defined as the likelihood that the software will fail when a request is made. For example, a POFOD of 0.00001 means that 1 every 100,000 requests may result in failure.

Another measurement in use is *Rate of Occurrence of Failure* (ROCOF), which is defined as the frequency of failures. For example, an ROCOF of 0.0001 means that there is likely 1 failure every 10,000 time units. Here, the time unit can be the clock hour or minute for nonstop real-time embedded systems or a transaction for transaction processing systems.

11.1.3 Improving Software Reliability

Real-time embedded systems typically require high reliability. Software reliability improvement techniques deal with the existence and manifestation of faults. Fault avoidance, fault removal, and fault tolerance are three major approaches to improve software reliability.

11.1.3.1 Fault Avoidance

Fault avoidance refers to a collection of practical techniques or rules of thumb in software design and development. Objected-oriented design and programming, formal modeling and verification, modularity, use of software components that are already proved to be fault-free, low coupling and high cohesion, and so on, are all examples of techniques of fault avoidance. In addition, documentation is an important aspect of fault avoidance, which is often ignored by many developers. Documentation includes requirements, design, analysis, assembly history, test cases, test results, change history, and so on. Documentation should be reviewed in each step of the system design and development process.

11.1.3.2 Fault Removal

Fault removal aims to detect the presence of faults and then to locate and remove them after the development stage is completed. It is performed through exhaustive and rigorous testing of the final application.

11.1.3.3 Fault Tolerance

Software is complex. It is not realistic to attempt to deliver a 100% bug-free software application, not to mention that most applications are extremely time-to-market-driven. Therefore, fault avoidance and fault removal are not sufficient to ensure high reliability. Fault tolerance is an approach that allows software to continue operation in spite of software failure. It proceeds with three steps: fault detection, damage assessment, and fault recovery or fault repair. Common practices for fault tolerance include N-Version Programming (NVP) and exception handling.

NVP is a method or process in software engineering where multiple functionally equivalent programs are independently generated from the same initial specifications. At runtime, all functionally equivalent programs are running in parallel on the same input to produce output. A voter is used to produce the correct output according to a specific voting scheme. NVD has been applied to software in switching trains, performing flight control computations on modern airliners.

Exception handling is the process of responding to the occurrence, during computation, of exceptions that change the normal flow of program execution. With exception handling, an exception breaks the normal flow of execution and executes a preregistered exception handler, which is normally implemented to handle the exception gracefully and resume the execution of the interrupted application.

11.1.3.4 Fault Recovery

Failure recovery is a process that involves restoring an erroneous state to an error-free state. The success of fault recovery depends on the detection of faults accurately and as early as possible. There are three classes of recovery procedures:

- Full recovery. It requires all the aspects of fault-tolerant computing.
- Degraded recovery. It is also referred to as graceful degradation. In this class of recovery, defective component is taken out of service.
- Safe shutdown.

Fault recovery can be performed with a *forward recovery* approach or *backward recovery* approach. In a forward recovery approach, procedures correct through continuation of normal processing. In a backward recovery approach, some redundant processes and state information are recorded with the progress of computation. The recovery proceeds by rolling back the interrupted processes to a point for which the correct information is available.

11.2 Software Aging and Rejuvenation

Software often exhibits an increasing failure rate and/or degraded performance over time due to accumulation of errors. This phenomenon is called *software aging*.

One major cause of software aging is memory leaks. A memory leak is a type of resource leak that occurs when a computer program incorrectly manages memory allocations in such a way that memory that is no longer needed is not released. Memory leaks are hard to detect. Memory leaks are a common error in programming, especially when using languages that have no built-in automatic garbage collection, such as C and C++.

Memory fragmentation is another factor of software aging. Memory fragmentation occurs when most of the memory is allocated in a large number of noncontiguous blocks or chunks – leaving a good percentage of total memory unallocated, but unusable for most typical scenarios. This results in out-of-memory exceptions or allocation errors (i.e., malloc returns null).

File descriptor leaks also contribute to software aging. It is important to ensure that files that are opened always get closed. Failing to close files will lead to I/O exceptions, a variety of failures on attempts to open properties files, sockets, and so on. File descriptor leaks can be detected through error messages in error log.

Software aging is inevitable. A proactive approach to handling the issue is to perform *software rejuvenation*. Software rejuvenation is the act of gracefully terminating an application and immediately restarting. Such a reboot of the system will remove the accumulated error conditions, free up system resources, clean the internal state of the software, flush the operating system kernel tables, and reinitialize the internal data structures.

Many users choose to wait until they see applications fail and then restart them. While this does not seem to be a bad idea, a preemptive rollback of continuously running applications would prevent failures in the future and minimize any collateral damage. There are two types of preemptive software rejuvenation. One is time-based or periodic. Businesses quite often do a regular reboot as part of a scheduled maintenance to help prevent aging-related Mandelbugs. For example, a database server might be rebooted every week during a period of low activity to reduce the probability of errors during busy times. It is reported that the telecommunication giant AT&T has implemented periodic software rejuvenation in the real-time billing system in the United States for most telephone exchanges. The second type of rejuvenation is prediction-based. In this approach, the date of the next software failure is predicted based on the previous failure data and a certain mathematical model. Rejuvenation is performed before the predicted date.

Rejuvenation events will cause the software application to be unavailable for the duration of the restart. In some situations, this will be included within a system's planned downtime allocation. In other situations, it will be unobservable and won't need to be accounted for.

11.3 Security

Similarly to traditional desktop and networked computing systems, real-time embedded systems face security threats as well. In fact, many real-time embedded systems are often required to store, access, or communicate data of a sensitive nature during their regular operation, making security a serious concern. The common information system security services, such as availability, confidentiality, authentication, data integrity, and nonrepudiation, are also important to embedded systems. The security goal is to protect sensitive data and/or resources from various kinds of attacks and malicious threats.

11.3.1 Challenges

Challenges in securing embedded systems come from limited processing power and memory capacity, restricted power budget, cost sensitivity, and open and specific operating environment. The limited processing and memory capacity of embedded systems make it impossible for their architectures to keep up with the continuously growing complexity of security mechanisms. Conventional security mechanisms tend to be conservative in their security guarantees, by adding a large number of messages and computational overhead, which not only presents a challenge to real-time task execution but also induces high-energy consumption. To face this challenge, energy-efficient security protocol execution is highly required. One solution consists of making the execution of employed cryptographic primitives more efficient through a combination of a new hardware and software optimization techniques.

Cost is also an issue in securing embedded systems. Embedded systems are often highly cost-sensitive; adding a few cents can make a big difference when a manufacturer builds millions of units of a product. Therefore, integrating top-level security is not always cost-effective for embedded systems because it mandates the use of more expensive hardware and software. Consequently, it is necessary to find a balance between the security requirement and cost in designing an embedded system.

11.3.2 Common Vulnerabilities

There is no 100% secure system though. Given sufficient time and resources, attackers can break any system. Therefore, designers' responsibility is to set up a reasonable security goal for a system based on the services it has to offer and all practical constraints it has to obey. The first step is to identify potential vulnerabilities and attacks for the system under consideration and then implement proper services to counter potential attacks.

Common vulnerabilities of embedded systems include the following:

Programming errors. Buffer overflows, unvalidated input, race condition, and insecure file operation are the most common programming errors that can lead to control flow attacks.

A buffer overflow occurs when an application attempts to write data past the boundary of a buffer. Buffer overflows can cause applications to crash, can compromise data, and can provide an attack vector for further privilege escalation to compromise the system on which the application is running. Technically, there are stack-based and heap-based buffer overflow exploitations.

Input validation is a fundamental measurement in securing an application. This is particularly true for web-based applications. Any input received by an application from an untrusted source is a potential target for attack. Examples of input from untrusted source include text input field, command input line, data read from an untrusted server over a network, audio, video, or graphics files provided by users, and so on.

Race condition issues were discussed in Chapter 6. The ATM example showed an undesired result when two people access the shared account in some particular orders. It is also vulnerability that attackers can take advantage of, if no mechanism that handles the race condition issue is implemented.

In some cases, opening or writing to a file in an insecure fashion can give attackers the opportunity to create a race condition. For example, you create a file with the write permission. Before you write to the file, an attacker changes the permission to read-only. When you write to the file, if you do not check the return code, you will not detect the fact that the file has been tampered with.

Access control problems. Access control in information system security is about the control of who is allowed to do what. Access control includes authorization, authentication, access approval, and audit. Authorization is the function of specifying access rights or privileges to resources. Authentication is the act of confirming the truth of an attribute of a single piece of data claimed true by an entity. Access is approved based on successful authentication. Many security vulnerabilities are created by the careless or improper use of access controls, which lead to attackers to gain more privileges that they should have. Of particular interest to attackers is the gaining of *root privileges*, which refers to having the unrestricted permission to perform any operation on the system. In general, the access control should be implemented fine-grained enough to satisfy the *principle of least privilege*, which limits access to the minimal level that allows normal functioning.

By the way, using weak passwords or hard-coded passwords in a device is obviously a bad idea, because such vulnerabilities make it possible for attackers to bypass access control mechanisms rather easily with minimal effort.

Improper data encryption. Encryption is the most fundamental mechanism to protect the confidentiality of the data stored in a computing system or transmitted via the Internet or other networks. Modern encryption algorithms play a vital role in the security assurance. In addition to confidentiality, encryption algorithms can be used to ensure data integrity. However, improper encryption can also lead to security vulnerability. Examples include using weak random number generators for generating cryptographic keys, trying to create your own encryption method, and attempting to implement a published encryption algorithm yourself.

11.3.3 Secure Software Design

Securing a real time embedded system requires a good software engineering practices and involves addressing the security concerns in every phase of the software development life cycle. It is a bad idea to perform security-related activities only as part of testing, because the after-the-fact technique usually results in a high number of issues discovered too late or not discovered at all. It is a far better practice to integrate activities across the software development life cycle to help build security in.

A risk assessment and vulnerability identification should be performed, and a cost-effective security goal should be established in the software requirements and design phase. Then, in the implementation phase, techniques that serve the security objectives of a particular system or product should be evaluated and implemented. After that, the secure code should be fully reviewed and thoroughly tested. In addition to regular testing that aims to remove bugs, security penetration testing is mandatory.

Penetration testing, also called *pen testing*, is the practice of testing a computer system, network, or Web application to find vulnerabilities that an attacker could exploit. The main objective of pen testing is to determine security weaknesses. Pen testing can be either performed manually or automated with software applications. The process includes gathering information about the target before the test, identifying the possible entry points, attempting to break in with fault injection, either virtually or for real, and reporting the findings back.

11.4 Safety

Many real-time embedded systems are safety-critical systems. Safety is the property of a system that it will not endanger human life or the environment. It is a measure of the freedom of a system from those conditions that can cause death, injury, occupational illness, damage to equipment or property, or damage to the environment.

Safety and reliability are two different properties of a system. Reliability is a measure of the rate of failures that render the system unusable, while safety is a measure of the absence of failures or conditions that would render the system dangerous. Safety is a combination of risk probability and severity of impact. Reliability can reduce the probability of an event of risk, but not its impact. A reliable system may not be safe, while a safe system may not be reliable. Imagine a car that is never able to run above 10 miles per hour; it is safe, but its transmission system is not reliable. If the car is functioning very well according to all functional specification, but its two real wheels fall apart when it runs over a small bump at a high speed, then it has a good reliability but is not safe. Therefore, safety should be ensured, so that if something happens, the impact is as low as possible.

Measurement of safety property involves how frequent and how severe are the dangers. Risk probability can be classified into extremely improbable, improbable, and probable. Risk severity is usually derived from system safety assessment, which is a means to identify and control hazardous consequences to be within acceptable risk.

The primary objective of all safety activities is to reduce or mitigate as many risks as possible, ensuring that the product does not have the potential to harm users. Developers of mission-critical embedded products, such as those in the medical device, railway, automotive, and avionics industries, to name just a few, need to have strategies to ensure the safety of their products. The rigorous standards and regulations that apply to safety-critical sectors have to be followed. In terms of software development, there are certain well-established quality assurance methods that help developers achieve functional safety. It is essential to address the safety issues of embedded products across the entire software development life cycle, starting with proper, accurate, and well-documented safety requirements. A mature process should be in place to track and manage the requirements, risks, and all quality assurance activities.

11.5   Power Conservation

Many real-time embedded systems are battery-operated and are thus constrained by limited power budget. Conventional system design metrics such as performance, size, weight, nonrecurring engineering cost, correctness, and testability are understandably important. Power consumption, however, is even more critical for embedded systems. Lower energy consumption means longer battery life and mission duration.

There are two different types of power consumption: *static* (also referred to as *standby*) and *active*. Static power consumption is mainly due to leakage. It increases with temperature and supply voltage. Since leakage is a natural phenomenon that comes with shrinking process technology, the only way to eliminate it is to shut that component down.

Active power consumption depends on chip activity. It increases with supply voltage, but not temperature. Several techniques have been developed for power conservation. *Dynamic voltage scaling* (DVS), which involves dynamically adjusting the voltage used in a computing component, is a well-known technique in power management for real-time embedded systems. DVS to increase voltage is known as overvolting; DVS to decrease voltage is known as undervolting. Undervolting is performed in order to conserve power.

DVS is often used in conjunction with *dynamic frequency scaling* (DFS), a technique in which the frequency of a microprocessor can be automatically adjusted "on the fly," either to conserve power or to reduce the amount of heat generated by the chip. Through DVS and DFS, quadratic energy savings can be achieved at the expense of linear performance loss. Thus, the execution of tasks can be slowed down in order to save energy, as long as the deadline constraints are not violated.

Other power conservation techniques include *dynamic power switching* (DPS), where software is used to switch between power modes based on system activity, and *adaptive voltage scaling* (AVS), a closed-loop, hardware and software cooperative strategy to maintain performance while using the minimum voltage needed based on the silicon process and temperature.

Power consumption is also affected by the way the software is designed. Display, wireless peripherals, USB, CPU utilization, and memory are among the key areas where software has an influence. Nowadays, many System-on-Chip (SoC) devices support power management by providing low power states, such as sleep, doze, and hibernate, and mechanisms for software developers to leverage them. In general, anything that can be done to reduce the number of clock cycles or the clock frequency needed to execute an algorithm can be applied to reducing power consumption, as long as a system can complete the work as required. Another example is with I/O. Since I/O buffers at the pin need to drive the current, we should minimize the traffic through major peripherals such as memory controllers and eliminate unnecessary data transfer in and out of the SoC.

Suggestions for Reading

Bohrbugs and Heisenbugs and their characteristics in software systems were first discussed in Ref. [1]. A more thorough classification and precise definitions of software faults were presented in Ref. [2]. The NVD technique was introduced in Ref. [3] and experimentally evaluated in Ref. [4]. Software aging and rejuvenation were discussed in Ref. [5]. Khelladi *et al.* [6] and Kocher *et al.* [7] are two good articles on embedded system security. DVS was first introduced in Ref. [8]. AbouGhazaleh *et al.* [9] presented an approach that a novel hybrid scheme uses DVS and compiler support to adjust the performance of embedded applications to reduce energy consumption.

References

1 Gray, J. (1985) Why do computers stop and what can be done about it? *Technical Report* 85.7, PN87614, Tandem Computers, Cupertino.
2 Grottke, M. and Trivedi, K.S. (2005) A classification of software faults. *Proceedings of the 16th International IEEE Symposium on Software Reliability Engineering*, pp. 4.19–4.20.
3 Chen, L. and Avizienis, A. (1978) N-Version Programming: A Fault-Tolerance Approach to Reliability of Software Operation. *Proceedings of the 8th IEEE International Fault-Tolerant Computing*.

4 Knight, J.C. and Leveson, N.G. (1986) An experimental evaluation of the assumption of independence in multi-version programming. *IEEE Transactions on Software Engineering*, **SE-12** (1), 96–109.

5 Castelli, V., Harper, R.E., Heidelberger, P. *et al.* (2001) Proactive management of software aging. *IBM Journal of Research and Development*, **45** (2), 311–332.

6 Khelladi, L., Challal, Y., Bouabdallah, A., and Badache, N. (2008) On security issues in embedded systems: challenges and solutions. *International Journal of Information and Computer Security*, **2** (2), 140–174.

7 Kocher, P., Lee,R., McGraw, G., Raghunathan,A. and Ravi, S. (2004) Security as a new dimension in embedded system design. *The 41st Design and Automation Conference*, San Diego, California, USA, June 7–11, 2004.

8 Pering, T. and Brodersen,R. (1998) Energy efficient voltage scheduling for real-time operating systems. *Proceedings of the 4th IEEE Real-Time Technology and Applications Symposium*, Denver, CO, June 1998.

9 AbouGhazaleh, N., Childers, B., Mosse, D., Melhem, R. and Craven, M. (2003) Energy management for real-time embedded applications with compiler support. *LCTES'03*, San Diego, California, USA, June 11–13 2003.

Index

Real-Time Embedded Systems, First Edition. Jiacun Wang.
© 2017 John Wiley & Sons, Inc. Published 2017 by John Wiley & Sons, Inc.

Printed and bound by CPI Group (UK) Ltd, Croydon, CR0 4YY

16/04/2025

14658592-0003